THE STORY OF CHANCE

WHAT'S LUCK GOT TO DO WITH IT?

MICHAEL ORKIN

innovativeink
PUBLISHING
A Division of Kendall Hunt

innovativeink PUBLISHING
A Division of Kendall Hunt

www.innovativeinkpublishing.com
Send all inquiries to:
4050 Westmark Drive
Dubuque, IA 52004-1840

Print ISBN: 979-8-3851-2252-3
eBook ISBN: 979-8-3851-2253-0

Published in the United States of America

CONTENTS

ACKNOWLEDGEMENTS

I'd like to thank the following people, who have been of great help during the course of this project: Angela Lampe and Katie Celarek at Innovative Ink, Bailee Bussan at Kendall Hunt, Gavin Huntley Fenner for numerous helpful talks on accident proneness, Katrina Uy for the Ethereum example in Chapter 22 and for proofreading the manuscript, Dan Anbar for inventing the term "stochastophobia," Philip Stark for pointing out the article he and the late David Freedman wrote on estimating the chance of major earthquakes, Robert Pisani for numerous discussions on chance in everyday life, and Yong Pin Yang, for her infinite patience while I was writing this book.

INTRODUCTION

"*Luck*: Success or failure brought by *chance* rather than through one's own actions."

"*Randomness*: The quality or state of lacking a pattern. Unpredictability."

"*Chance*: The possibility or likelihood of something happening. The occurrence and development of events in the absence of any obvious design."

"*Probability*: The likelihood of something happening."

Oxford Languages Dictionary

In the beginning, there was earth, wind, fire, gravity, magnetism and… chance. And then came life. After a while, evolution, with the support of genetics (chance mutations), caused living organisms to become smarter and, for some (the fittest), better able to survive. More time passed until, finally, here we are.

We got lucky. We won the genetic lottery. Or not. If you're bacteria in a pond or a fruit bat, better luck next time (no offense intended). You can't see randomness, but it's a thing, and you can experience its effects. There's a 50% chance that a coin will land heads or tails, but the result is unpredictable from toss to toss. Same for rolling a pair of dice. Same for buying a lottery ticket. Same for what will happen to you next year, which is analogous to a lottery except more complicated and hopefully more favorable. In the world of quantum mechanics, randomness governs the behavior of subatomic particles. The casino games of roulette and craps are determined by luck. No skill involved. Simulation software models randomness ranging from the weather to blackjack, processes too complicated to understand using the classical mathematics that has been around since Pythagoras.

What is randomness? How did the study of probability, which is how we measure randomness, get started? What is the *law of averages*? What is *expected value* and how do casino games illustrate this? Why does a lucky lottery jackpot winner need unlucky losers? How are the occurrences of "black swans" explained by the *opportunity* for such things to happen. What does it mean for a gambler to *cherry-pick*? What's the difference between the chance of winning the lottery and the chance of being killed by a bee sting? What is *apophenia*? What is *synchronicity* and why did Jung invent the term? Why can you have a winning strategy in blackjack but not in craps?

A TV pundit has predicted every stock market crash for the past thirty years. How is that possible? What's the double-up strategy in roulette? What's the Karaoke strategy? Can an octopus successfully pick winners of football games? Why does the bell-shaped curve show up so often? What is the "margin of error" you see in opinion polls? What's a p-value? How can you tell whether or not a fortune teller has psychic powers? How can you tell the difference between luck and skill? Is sports betting skill-based? Was Humpty Dumpty accident prone? What's the easiest way to predict the value of one variable based on another variable? Does correlation mean causation? Why does the Federal Reserve fail to understand basic statistical facts? What's Bayes' Rule and how can it be used to update the probability of having a rare disease based on a diagnostic test? What is resemblance stereotyping and how can conditional probability help understand it? How did the former crypto king con himself? Why are lotteries sucker bets? I'll discuss all these things and more, and I'll try to answer the question, "What's luck got to do with it?"

At the end of each chapter there is a "Bottom Line" paragraph summarizing the main concepts of the chapter.

There's the inevitable math, but you can skim over math that seems too complex and still understand the concepts. I've included a couple chapters at the end of the book to explain two probability distributions. I use * for multiplication and sometimes ^ for exponentiation, same as in spreadsheets.

CHAPTER 1

THE KING'S LOTTERY (RANDOM ALTRUISM)

Once upon a time, there lived a king who amassed great wealth by plundering and pillaging and also by extorting a hefty tax from anyone doing business in the kingdom. Yet, although the king was ruthless and greedy, he announced one day that he wanted to share his wealth with his citizens by giving it away at random. He called his plan "random altruism." To accomplish this, the king created a lottery.

A ticket to the king's lottery cost a gold coin, the currency of the day. Every coin had a picture of the king on it. Once a ticket was purchased, a player would pick twenty numbers between 1 and 80. There was a drawing to pick the winning ticket every Saturday night. To discourage gambling addiction and to lower the chance of someone getting lucky, each citizen could buy only one ticket per drawing. For the drawing, there was a large urn in the king's palace. The urn contained 80 balls, numbered from 1 to 80. With appropriate fanfare, the king himself randomly selected twenty of these balls. The numbers on the balls selected by the king were the winning numbers. If all 20 numbers on someone's ticket matched the 20 winning numbers, then that person would win the king's lottery. If more than one person picked the winning numbers, the winners would split the pot. There were no secondary prizes. Winners would get a trunkful of the king's loot: Gold bullion, jewelry, coins, precious stones, deeds to land, and other valuables, worth about $1 billion in today's dollars.

There was no internet in those days. No phones either. Although people heard rumors about a winner in a faraway town or in a city on the other side of the mountains, these were just rumors. In fact, nobody ever won. The king, who had some understanding of probability, knew that nobody would win. He set it up that way. He wasn't cheating. As advertised, the winning numbers were randomly selected. Although it was mathematically possible to win the King's lottery, the chance of winning is so low, about 1 in 3.5 quintillion, that for all practical purposes, nobody would ever win, at least not in the king's lifetime (which was tragically shortened by an infection from a mouse bite). There's simply not enough opportunity. For example if everyone on earth today (8 billion people) bought a ticket for each weekly drawing of the king's lottery, there would be an average of one winner every 9 million years. In the king's day, there were less than a million people on earth, so nobody was going to win. In addition to being a plunderer and pillager, the king was also a con man. Random altruism was a farce.

Even though most citizens realized that they probably wouldn't win, they didn't care and played the king's lottery anyway. They did this to show their loyalty to the king and to bond with their fellow citizens who were also showing their loyalty. Saturday night became a social event, a ritual in towns and villages across the kingdom. Once the winning numbers were known, even people in the most remote provinces would soon learn that they lost from couriers.

The king, who craved publicity, loved the lottery drawings every Saturday night, when all his subjects would come together in town squares and meeting halls and arenas and amphitheaters to participate in lottery festivities. Food and drink were plentiful. Not only did the lottery give all the citizens a sense of camaraderie and an opportunity to show their loyalty to the king, the king made a boatload of money from ticket sales.

It was all a scam, verified by the laws of chance. One night, after a particularly lucrative lottery drawing (more tickets sold than usual because of a full moon), the king's accountant remarked, "The probability of someone winning is less than the probability of a huge asteroid falling out of the sky and killing everyone on earth." The king replied, "Suckers have no business with money, anyway." Many centuries later, this comment was used by Canada Bill Jones, a 19th century con man and riverboat gambler.

Bottom Line

Some events, like winning the jackpot in the King's Lottery, have such low probabilities that they will most likely never happen, at least in our lifetime.

CHAPTER 2

THE MEGA MILLIONS LOTTERY

"If it wasn't for bad luck, I wouldn't have no luck at all."
"Born Under a Bad Sign," by Albert King.

"All Lottery Tickets are Created Equal" – patriotic gambler

alisafarov/Shutterstock.com

The Mega Millions lottery is played in 45 states, the U.S. Virgin Islands, and the District of Columbia. It's not played online. The jackpot prize is always in the millions. Despite various popular lottery "strategies," like using lucky numbers, birthdays, anniversaries, astrology, numbers that are "hot," numbers that are "cold," numbers that are "due," and so on, there is no skill in buying lottery tickets. It's all random, because jackpot winners are determined by picking numbers at random. As such, lottery winners are just lucky. No skill is involved. You can't have a successful strategy predicting something that's unpredictable. But maybe the equipment is broken. Maybe the game is rigged. Maybe the lottery is a deep-state hoax. Maybe you think that a configuration of numbers you saw in your oatmeal this morning will be jackpot winners (in which case, you have *apophenia*, the tendency to perceive a connection between unrelated or random things). Or maybe there is a connection. Maybe there's a force that enables some people to pick winning lottery numbers. Maybe the famous 20th century psychiatrist and author Carl Jung was right and there is a noncausal connection, called "synchronicity," between your oatmeal and the lottery jackpot. Spoiler alert: No such evidence has ever been found, not by Jung or anyone else. Although if a lucky jackpot winner thought about it enough, they could undoubtedly connect winning to some event. People like to provide explanations for the occurrence of unlikely events, even when the only explanation is chance. Jung's "synchronicity" has been around for thousands of years. It's usually called "luck."

It costs $2 to buy a Mega Millions ticket. You pick 5 numbers from 1 to 70 and a "Mega-number" from 1 to 25. At the lottery drawing, 5 winning numbers are randomly selected from the numbers 1 to 70 along with one winning Mega-number from 1 to 25. If all 5 of your selected numbers along with your Mega-number match the winning numbers, you win the jackpot. The chance of winning the jackpot is 1 in 302,575,350. It's hard to put small probabilities in perspective, although analogies sometimes help. If you buy fifty Mega Millions tickets each week, you will win the jackpot on an average of once every 116,000 years. If you drive a mile every time you buy a ticket, you will drive an average distance of 633 round trips to the moon before you win the jackpot. If you have one friend in Canada and select someone at random from the entire population of Canada (38 million at the time of this writing), you are about 8 times more likely to select your friend than you are to win the Mega Millions jackpot. If nobody wins the Mega Millions jackpot, the prize money is rolled over to the next drawing. This continues until somebody wins the accumulated jackpot.

Opportunity

Even though it's almost impossible for a particular ticket to be a winner of the Mega Millions jackpot, there are often winners. Why? It's because of the *opportunity* afforded by the sales of millions of tickets. In the Mega Millions lottery, *you* may not win the jackpot, but with enough tickets sold, there is a good chance that *someone* will win, as follows: Given enough *opportunity* weird events will happen just due to chance. Statisticians Percy Diaconis and Frederick Mosteller called a variation of this phenomenon the "Law of Truly Large Numbers." In street vernacular, "S**t happens, just due to chance."

In this case, "opportunity" means many tickets being sold for the Mega Millions drawing. This is true of unlikely events in general. If you toss a coin 20 times, the chance is very low, about 1 in a million, that you will get heads on every toss, but if a million people each toss a coin 20 times, someone will likely get lucky and get heads on every toss. The world was full of white swans until someone noticed a black swan in Australia. The unlikely black swan became the icon for Nassim Taleb's bestselling book of the same name. Part of Taleb's definition of a "black swan" event was that it be a very unlikely, meaningful event. The same is true of lottery jackpot winners. You don't hear about the losers, only the person who won the jackpot. Although it's very unlikely that a particular person will be a jackpot winner, if enough people buy tickets, it's not so unlikely that *someone* will win. The same was not true of the King's Lottery because the chance of winning the jackpot was so low that even if everyone on Earth bought a ticket, there wasn't enough opportunity for a winner.

Multiple Ticket Purchasers

Many die-hard lottery players don't buy one ticket for a drawing, they buy many tickets. It's true that the more (different) tickets you buy, the greater your chance of winning. For example, if you buy ten tickets, your chance of winning is ten times greater than if you buy one ticket, but that means that your chance of winning the Mega Millions jackpot equals about 1 in 30 million compared to about 1 in 303 million, still a very low probability. Don't forget, a ticket costs $2, so if you buy ten tickets your cost is $20.

Suppose you buy every possible Mega Millions ticket combination, all 302,575,350 of them. Doesn't that guarantee that you'll win the jackpot? Yes, it does, but there are problems. As of this writing, the largest jackpot in Mega Millions history was $1.602 billion, won by a lottery player in Neptune Beach, Florida, on August 8, 2023.

Suppose you had purchased all possible ticket combinations for that drawing. At $2 per ticket, this would have cost $605,150,700. Note that neither the price of buying a ticket nor the probability of a particular ticket winning the jackpot changes when the jackpot gets high. Since you purchased every possible ticket combination, you guarantee that you will win the jackpot. Unfortunately for you, there was another winner, the person from Neptune Beach, so you would have had to split the pot, and your jackpot winnings would have been reduced to $801,000,000 (half of $1.602 billion). You then would have had to pay about 1/3 of this in taxes, further reducing your winnings to about $534,000,000. Already you would have lost about $71 million, and this doesn't consider the fact that jackpot winnings are spread out over 20 years.

So, even though you guarantee winning the jackpot, buying every ticket would have been a bust. In addition, if you fill out 3 tickets per minute, 24 hours per day, 7 days per week, it will take you about 192 years to fill out all the ticket combinations. This is unlike the racetrack, where you can "box" all combinations of certain bets and make a combination bet with a single ticket, thereby losing a bundle in the time it takes to cry in your beer. Mega Millions software doesn't allow you to bet on all ticket combinations. You have to write out every ticket individually and, in the great American tradition, work hard for your losses.

A Glimmer of Hope?

Although your chance of winning the jackpot is the same no matter how large the prize and no matter how many people buy tickets, there is something you can do to possibly make your winnings larger than if you pick numbers at random: Pick numbers unpopular with the betting public. If you choose numbers that nobody else picks, your chances of winning are still the same, 1 in 302,575,350 for each ticket purchased, but if you happen to be so lucky as to win, you are less likely to split the pot with another winner if there is one. The numbers that are most popular with the betting public are lucky numbers such as 7 and 11, and birthday days (days of the month), numbers from 1 through 31, so don't select these numbers when you buy a ticket. However, if enough people follow this strategy, the numbers I just recommended will become popular and you'll have to switch to other numbers. And so on and so on. Whether or not you split the pot with someone if you win might not matter because your chance of winning is so low to begin with. Still, small changes to betting strategies can sometimes make big differences in hypothetical winnings.

The Station

Many years ago, there was a restaurant in Berkeley, California, called the Station, so named because it was located next to the West Berkeley Amtrak station. In the front of the restaurant was a sports bar, where a local bookie would take bets on sporting events and watch the games on TV. I would hang out at the Station with Les, my friend from college who was the bookie's attorney. The bookie was known as the "Plumber." I figured that he was an underworld character who had this nickname because when his customers didn't pay their debts, he would dunk their heads in a toilet or hit them with a plunger, or something terrible that involved plumbing. One day, I got up the nerve to ask him why he was called the plumber. He said, without taking his eyes off

hxdyl/Shutterstock.com

the large screen TV mounted above the bar, "Because I am a plumber. Being a bookie is what I do after work." One evening, the day before a lottery drawing, Les and I were at The Station. When I was returning from the restroom, I glanced in the kitchen and saw kitchen workers sitting around a large table filling out forms. When I asked the Plumber what they were doing, he said that the day before lottery drawings, the chef had all the kitchen workers fill out lottery tickets in hopes of winning. They would fill out thousands of tickets for every drawing. Although they won many small prizes, they never won the jackpot.

At first glance, it seems like both the Mega Millions Lottery and the King's lottery have such low probabilities of winning, that if enough tickets are sold to provide the opportunity for a winner in one game, then there should be the same type of opportunity for a winner in the other. The chance of winning the Mega Millions Lottery jackpot is about 1 in 303 million, whereas the chance of winning the King's Lottery jackpot was about 1 in 3.5 quintillion, making the chance of winning the Mega Millions jackpot 11.7 billion times more likely than winning the King's lottery. Since one interpretation of probability is about the frequency of events happening over time, and since humans generally don't live longer than 100 years, the difference in these probabilities is important, as the king and his advisors knew when they advocated random altruism.

A Tragic Story

According to a Reuters article, on July 6, 2021, "A British teenager was found guilty on Tuesday of murdering two sisters, as part of a pact he believed he had made with demonic forces to kill at least six people every six months in exchange for a future lottery win... he wrote that he hoped his reward would be to 'win the Mega Millions Super Jackpot', and police found three lottery tickets he had bought after the murders." The article doesn't say whether the tickets found by the police were winners. Hopefully, making a pact with demonic forces doesn't increase your chance of winning.

What if There's Not Enough Opportunity?

The chance of winning the King's lottery was so low that there was not enough opportunity for a winner even if everyone in the kingdom bought a ticket. The same isn't true for the Mega Millions lottery, which has frequent jackpot winners because many tickets are sold, and because the probability of winning the Mega Millions jackpot is much higher than the chance of winning the King's lottery. But what if somebody wins the Mega Millions jackpot and it turns out that no other tickets were purchased for the drawing? Is this person lucky? You might think so, because why should the number of other tickets sold affect the winner?

The probability of winning the jackpot is so low, if only one ticket is sold and that ticket is a jackpot winner, then you might conclude that the winner isn't just lucky and look for a more plausible explanation, like cheating or faulty equipment. Or maybe this person is a "false positive" in the sense that the result looks suspicious but really it's just due to chance. Oddly, the number of tickets purchased affects whether a winner is lucky or maybe cheating. To paraphrase the Beatles song, Mega Millions jackpot winners get by with a little help from their friends (the losers). In this example, since no other tickets were sold, there are no losers to "help" the winner but only a single ticket holder who mysteriously won. Here's a big question: How do you decide statistically if a jackpot winner is just lucky or if something else is going on? The answer is to use "p-values."

The important thing here is not the probability that a particular person wins the jackpot, but the probability that *someone* wins the jackpot. If only one person buys a ticket, these two probabilities are the same, 1 in 302,575,350. As more tickets are purchased, the probability that a particular person wins the jackpot remains unchanged, but the probability that *someone* wins the jackpot increases. The winner is a particular person but the reason there are lucky winners is because there are enough tickets sold so that *someone* wins. This is the secret of why weird, unlikely things happen just due to chance.

The statistical way to decide if a jackpot winner is lucky or cheating (or some other plausible alternative to luck) is with a *p-value*, the probability that an event occurred just due to chance. There's no skill in buying lottery tickets. If a jackpot winner isn't lucky, the alternative is cheating, broken equipment, or some plausible alternative to chance. Why should the lack of losers have anything to do with why someone was the winner? Didn't the winner get lucky and win the jackpot regardless of how many other tickets were sold? Not really.

Opportunity means lots of tickets are sold. We can compute the chance that *someone* will be a jackpot winner. The chance that Mr. Johnson wins the jackpot always remains the same, but the chance that *someone* wins the jackpot increases as more tickets are sold. How many tickets need to be sold to have statistical evidence for a lucky winner if there is one? In statistics, you typically attribute a result to luck if the p-value exceeds 0.05. Using this cutoff point, we need to find the minimum number of tickets, k, so that the probability that *some* ticket wins the jackpot just by chance exceeds 0.05.

How much opportunity is needed to conclude that a jackpot winner is lucky?

The probability that a particular person wins the jackpot equals 1/302,575,350, and so the probability that a particular person doesn't win the jackpot equals 1 - P(win) = 302,575,349/302,575,350. Since the results of each ticket (win the jackpot or not win the jackpot) are independent, it follows that the probability of no jackpot winners in a drawing in which k tickets are sold equals (302,575,349 / 302,575,350)^k. (It's okay to skip the math.) Subtract this from 1 to get the probability that there is at least one winning ticket, set this equal to 0.05, and solve for k. The result: k= 15,520,087. At least 15.5 million Mega Millions tickets must be sold to conclude with statistical confidence that if someone wins the jackpot it's because they're lucky (This includes rollover tickets from drawings when there were no winners). Otherwise, there's a problem.

The key idea is that for unlikely events to happen due to chance, opportunity is necessary. If you see the occurrence of an unlikely event, and there isn't enough opportunity for the event to happen just by chance, then perhaps the event didn't occur by chance. Without losers, there can be no lucky winners. In the world of chance, success is intertwined with failure. You can't have one without the other. Next time you look around, marvel at the beautiful world we live in and give thanks to the unlucky losers who made it all possible.

Bottom Line

The reason there are Mega Millions jackpot winners, even though the chance of winning is so low, is the opportunity that comes from millions of ticket sales. Given enough opportunity unlikely things happen just due to chance. As we saw with the King's Lottery, this isn't always true, since sometimes the probability of success is so low that there's not enough opportunity for the event to happen in a reasonable amount of time.

CHAPTER 3

SIR ISAAC NEWTON AND THE GIANT CLOCK

Sir Isaac Newton, one of the founders of classical physics, thought the world worked, metaphorically, like a giant clock. Nothing was random. No events were due to chance. Scientists had only to figure out how different parts of the giant clock worked and we would understand everything. Unfortunately, or fortunately, depending on your point of view, things didn't work out that way. Even though it took a few hundred years, by the beginning of the twentieth century, Einstein's theory of relativity and the emergence of quantum mechanics showed that the world couldn't be modeled very well by a giant clock. In fact, Einstein's work on the space-time continuum showed that even a giant clock couldn't be modeled very well by a giant clock, which now began to look like a Dali painting.

Luckily, as statisticians and gamblers like to say, probability theory and the beginnings of modern statistics were also developed in the same few hundred years, so the scientific world had tools to help deal with the partial demise of classical physics and the understanding of chance.

I say partial demise because classical physics still works well in many situations and is good at explaining "big things" as opposed to subatomic things. It turns out, however, that even with big things, chance can interfere with the giant clock envisioned by Newton and his contemporaries. Obvious sources of randomness are dice, coins, cards, and other devices used to generate random numbers and prophecies, along with genetics, which manifests itself in all forms of life. Still, when we sit down, we don't worry about our chair suddenly catching on fire, however, even though the probability of this weird event is very low, it's not zero. For example, it happened here: "Ashley Furniture faces civil lawsuit after recliner catches fire, gives man severe burns." (https://www.fox32chicago.com/news/ashley-furniture-faces-civil-lawsuit-after-recliner-catches-fire-gives-man-severe-burns) (To be fair, the chairwas a recliner with an electrical component, although not an "electric chair.") When we take a hike in the woods, we don't expect to be hit by a random airplane. The probability isn't zero, however, and, guess what, it happened: (https://www.msn.com/en-us/news/us/pair-out-for-walk-hit-by-crashing-plane-in-indiana/vi-BB1kU5wN?cvid=0694ab335a13404b98cb6bfa223ec4f0&ocid=winp2fptaskbar&ei=24&sc=shoreline#details)

Things that we assume won't happen because they're so unlikely, like chairs spontaneously bursting into flame and planes falling on hikers in the woods, occasionally happen. As we saw with the King's Lottery, even though it's a random event, winning the jackpot almost surely won't happen, at least not in our lifetime, because the probability is so low. Our life is filled with low probability events, most of which never happen in our lifetime, but which add randomness (uncertainty) to the world, and when you put them all together, their combined effect causes some to happen. Mosteller and DIaconis say that "We are swimming in an ocean of coincidences." When we walk across a room, we don't worry about having a heart attack, but there's a non-zero chance that it will happen (more likely for some than others.) This makes the macro world less Newtonian and closer to the subatomic, quantum world. The ground beneath us is firm at the moment, but it will break as the result of a giant, unpredictable earthquake sometime in the distant future, maybe sooner if you live in California. At an unpredictable time after that, the sun will burn out, but this will be long after we're gone. We don't worry about pigs learning how to fly, and I would say that the probability of this is zero for all practical purposes, although we don't know what strange paths evolution will take. In the short term, we can count on a reasonably stable existence, but a giant clock is not a very good metaphor for our physical world.

The Rashomon Effect in Real Life

Some people don't like the idea of things happening at random or even the potential for something to happen at random. They insist that everything can be explained and is therefore predictable and non-random. The fact that some things can't be explained (except by randomness) threatens their well-constructed house of cards. A friend of mine, Dan Anbar, calls this "stochastophobia," the fear of randomness. It's related to, but different from "apophenia," the tendency to perceive a connection or meaningful pattern, sometimes foreboding, between unrelated or random things. There are still a few diehards who want to return to the glory days, when Newtonian physics presumably explained everything, and who believe that there's no such thing as randomness. The proof of this, some of them say, is that you can look in the past and reconstruct in detail exactly how and why something happened, thus providing a reason for everything and, in the process, eliminating randomness. This is a low bar for predictability. The usual definition regards predicting something that has not yet happened.

It's not easy to explain the past. Consider the "Rashomon effect," which refers to a 1950 movie made by the Japanese filmmaker Akira Kurosawa, based on a story of the same name. In Rashomon, the killing of a samurai and the assault of his bride are witnessed by 4 different people: the bandit, the bride, the samurai's ghost, and a woodcutter, who all give conflicting, contradictory accounts of the same crime. The same is true for many situations, ranging from eyewitness accounts of crimes to scientific experiments. Even recounting the past is unpredictable in a sense.

In "The Black Swan," Nassim Nicholas Taleb mentions Pierre-Daniel Huet, a 17th century Catholic bishop and skeptic who wrote a book called "Philosophical Treatise on the Weakness of the Human Mind," in which he says that "any event can have an infinity of possible causes." Perhaps not an infinity but certainly a lot of possible causes, making it difficult to explain precisely why something happened (see Chapter 18, "The Nowhere Man") This doesn't mean that we live in a random cauldron, but maybe Nobel prize winning physicist Neils Bohr's famous saying, "Prediction is very difficult, especially if it's about the future" should be modified to be "Prediction is very difficult, whether it's about the future or the past."

Chaotic Systems Model Reality but are Different From the Reality They Model

We try to make sense of the world by observing processes and patterns and finding ways to measure things. We look for connections and try to replicate experiments. Quantum physicists have discovered that measuring a subatomic particle has a big effect on the particle, not just because of clumsy measuring devices, but due to the measurement itself. In the subatomic world, there's no such thing as a neutral "observer", a fly on the wall who can see what's going on without interfering with what's being observed.

Even though the experimenter (observer) doesn't necessarily interfere with nature in the non-quantum world, sometimes a model of some aspect of nature gets confused with nature itself. In chaos theory, "chaotic" systems like the weather have a property in which small changes in initial conditions of a model for predicting the weather can lead to unpredictable variation in a future state of the system. This makes weather prediction more than a few days in advance difficult when using these methods.

This phenomenon was discovered and made famous by the mathematician and meteorologist Edward Lorenz in the early 1960's. Lorenz's discovery was figuratively called the "butterfly effect." In 1972 Lorenz gave a talk at conference of the American Association for the Advancement of Science. His paper was called "Predictability: Does the Flap of a Butterfly's Wings in Brazil Set Off a Tornado in Texas?" Lorenz's model of the weather involved certain types of mathematical functions that were deterministic, not random, but for which an approximation to the initial terms of such functions could lead to grossly inaccurate predictions of future states of the system. The discovery of such systems showed that unpredictable systems could be based on deterministic functions and weren't random in the same sense as lotteries and dice. The problem was that some researchers subsequently treated Lorenz's model of the weather as if it were the weather itself. This confuses the model with reality. (The model is its own reality, residing on a computer.) The weather is unpredictable, for sure, but the full picture of the weather is not a mathematical model, even an elegant and successful one like Lorenz's. Today's weather forecasting involves combined simulations of numerous models.

Time

Time is a human construct. In Kurt Vonnegut's classic novel, "Slaughterhouse 5," Billy Pilgrim is a character who's "unstuck in time," which means that he moves through time in a nonlinear way. Einstein's theory of relativity shows that time and space are warped, in a sense, so that in the big picture time isn't linear. It's possible to envision a world where you are removed from time entirely and projected onto space. Like Billy Pilgrim, you would be unstuck in time and just watch things unfold continuously or however things unfold without time. And without time, what is the use of a giant clock?

Probability

The way to measure randomness, or chance, is with probability. For example, we can toss a coin and say that since there are two possible outcomes, "heads" or "tails," each of which has the same chance of occurring, then the probability is 50%, or 50-50, or 0.5 that the coin comes up heads. One interpretation of this is that in repeated tosses, the coin will come up heads about 50% of the time. If we roll a die, since there are 6 faces, with 1 through 6 dots on them, we say that the chance that "3" comes up equals 1/6, with the interpretation being that in repeated rolls of the die, "3" will come up about 1/6 of the time. This type of probability is supported by strong mathematical results and provides a way to measure the likelihood of many random events. Unfortunately, this nice probability package is not always available. For example, what's the probability that a stock goes up by at least 3% tomorrow? What's the probability of a major earthquake in California next year? What's the probability that you'll get stung by a bee this summer? These situations, although random, are more complicated than tossing coins or rolling dice and must be analyzed in different ways.

Bottom Line

Randomness (chance) is everywhere, but we lead our lives without worrying about it, unless we are in a casino or otherwise engaged in risky activities.

CHAPTER 4

DICE, CARDS, COINS, AND LOTS

DICE

Devices that generate random numbers or symbols have been around since ancient times and have been used to play games and get divinations. When you play a dice game, you're also getting a divination and leaving your fate (win or lose) up to chance. You're also taking a random sample in the sense that you're randomly sampling the dice to determine an outcome. Based on archeological findings, dice shaped from the knuckle bones of hoofed animals were used for fortune telling and game playing since before recorded history. According to Wikipedia,

"Perhaps the oldest known dice were excavated as part of a backgammon-like game set at the Burnt City, an archeological site in south-eastern Iran, estimated to be from between 2800 and 2500 BCE. Bone dice from Skara Brae, Scotland have been dated to 3100–2400 BCE."

These discoveries make the 16th century dice game in Figure 1 seem modern, as the dice in Figure 2. Fortune telling is like a game. You try to accurately predict the future based on clues that are randomly uncovered by dice or some other chance device. I will discuss the modern dice game of craps later, along with other games played in the temple of chance, otherwise known as a casino. I will also discuss fortune telling. The most popular fortune telling system of ancient times, still in use today, one that uses random numbers generated by coins or yarrow stalks, is the I Ching.

Morphart Creation/Shutterstock.com

FIGURE 1 - 16th century dice game

riekephotos/Shutterstock.com

FIGURE 2 - Ancient Dice

The I Ching

One of the oldest methods for using chance to get a divination was in China, in the period 1,000- 750 B.C., when yarrow stalks were manipulated or coins were tossed to get a reading from the I Ching, or "Book of Changes." Using randomness assured a divination that was not in the user's conscious control. There are many stories and myths concerning the origins of the I Ching divination system. In one story, Fu Hsi, the first emperor of China (circa 2,500 B.C.), observed a turtle emerging from the Yellow River. Fu Hsi had an epiphany about the significance of the eight symbols he saw on the turtle's shell and how these patterns reflected the movement of energy on Earth.

Turtles were revered in ancient China and elsewhere. Following the accomplishments of Fu Hsi, there is evidence that patterns on tortoise shells that were heated over a flame until they cracked were used for divinations. In some cases, the shells were organized and marked with interpretations and stored for reference, perhaps establishing the world's first database (or turtlebase). A "world-bearing turtle" (see Figure 3) appears in Hindu mythology, Chinese mythology, and the mythologies of some indigenous peoples.

A story involving tortoise shells describes an ancient group of women fortune tellers who read the shells of turtles and who, according to legend, became the royalty of the Shang Dynasty. There's even a movie based on the book called, "Turtles All the Way Down," which refers to the following joke: When a child is told that the earth is held up by a giant turtle (see Figure 3), he asks, "What's holding up the turtle?" The reply is, "Another turtle." The child asks, "What's holding up *that* turtle?" The reply is, "Turtles all the way down."

A detailed account of the history surrounding the development of the I Ching can be found on divination.com as well as in the existing books about the I Ching, such as the comprehensive Wilhelm/Baynes book published by Princeton University Press.

The result of I Ching coin tosses and yarrow stalk manipulations is a hexagram. Figure 4 contains the 64 possible I Ching hexagrams, with each hexagram having 6 lines, one above the other. Each line represents "yang" (solid line) or "yin" (line with gap in the middle) For example, the line "---" is yang and the line "- -" is ying. Each line is randomly chosen according to a rather complex formula by tossing the three I Ching coins shown in Figure 5 (yin or yang instead of heads or tails). After you toss the three coins a number of times, one of the 64 hexagrams is generated (see Figure 4).

MoreVector/Shutterstock.com

FIGURE 3 World-bearing turtle holding up elephants and the world

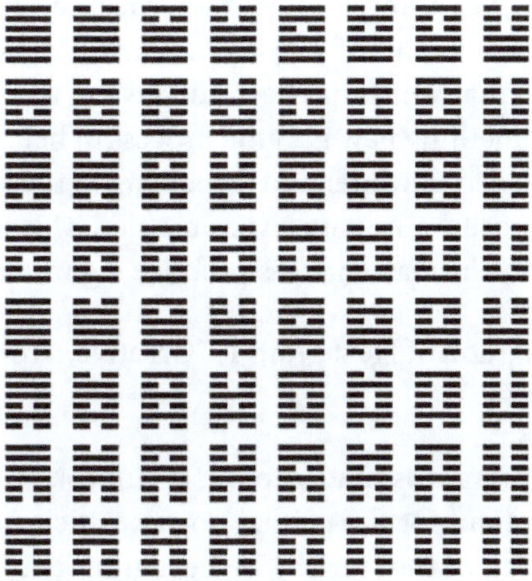

Lena Pronne/Shutterstock.com

FIGURE 4 - The 64 I Ching hexagrams

ArtWell/Shutterstock.com

FIGURE 5 - I Ching coins, yin and yang in place of heads and tails

A hexagram imparts a nugget of wisdom, an understanding of how the seeker fits into the cosmos, interpreted, perhaps, by a professional oracle, or maybe the seeker is making a DIY divination, which is easy to do using an I Ching app. Here is the first paragraph of a translation of the meaning of Hexagram 59, "Dispersing," from divination.com:

> "Nothing survives forever—not even rocks or the most rigid of structures. Likewise, all of our obstacles will dissolve in time too. The erosion of that which has been solid is not a bad thing; in fact, it means that something new is in the process of being created. Picture ice floes, hardened in winter and dispersing in the warmth of spring; when the ice melts, a mighty river emerges. This image points to a situation in which small changes produce big results over time."

Here's the first paragraph of Hexagram 1, "Creative," from the Wilhelm/Baynes book:

> "The first hexagram is made up of six unbroken lines. These unbroken lines stand for the primal power, which is light-giving, active, strong, and of the spirit. The hexagram is consistently strong in character, and since it is without weakness, its essence is power

> or energy. Its image is heaven. Its energy is represented as unrestricted by any fixed conditions in space and is therefore conceived of as motion. Thus, the hexagram includes also the power of time and the power of persisting in time, that is, of duration."

This is a simplification of the ritual involved in doing an I Ching divination. The I Ching book has translations into many languages. It is widely read and consulted around the world. In the 1960's, the I Ching was popular during the psychedelic movement in Berkeley and San Francisco, popularized by people such as Alan Watts, the Zen scholar and psychedelic explorer who had a long-running show on KPFA, the public radio station in Berkeley, California. Watts made the connection between the yin and yang of the I Ching and the 1's and 0's that comprise computer code. He might have done this after taking LSD.

It can be argued that randomness itself is symbolized by the I Ching. When something happens at random, it changes unpredictably, just like the I Ching divination changes according to the toss of the yin-yang coins or yarrow stalks. The process of consulting the I Ching not only uses randomness but is a metaphor for randomness and demonstrates the reverence paid by the ancient oracles for unpredictable change.

Today, the need for a professional oracle or I Ching interpreter has been diminished by I Ching apps on both the iPhone and Android. An I Ching app uses a random number generator in its software to toss simulated I Ching coins and generate a hexagram. It then searches through the app's I Ching database to find an interpretation for the chosen hexagram. What better way to interpret I Ching hexagrams originally found by tossing yin-yang coins than by using the computer's inherent yin-yang structure?

Hmm – I Ching, iPhone, iPad, seems like they go together. Steve Jobs was known to have studied the I Ching when he dropped out of college, before he started Apple. See https://miraprabhu.wordpress.com/2014/11/22/john-nicholas-stoodley-i-ching/

Cards

Just like dice and coins, cards have been used to generate chance outcomes for both games and fortune telling since ancient times. According to Wikipedia and other sources, playing cards were probably invented in China in the 9th century during the Tang dynasty when woodblock printing was developed. Other cultures followed suit, so to speak, as card games were introduced in Egypt, and then in Spain and Italy before moving to the rest of Europe in the 14th century. During this period, tarot cards were first introduced, possibly by wandering gypsies, as playing cards in Europe (see Figure 6). Sometime later, they were used for divination, called cartomancy. From Wikipedia:

> "The earliest evidence of a tarot deck used for cartomancy comes from an anonymous manuscript from around 1750 which documents rudimentary divinatory meanings for the cards of the Tarocco Bolognese. The popularization of esoteric tarot started with Antoine Court and Jean-Baptiste Alliette (Etteilla) in Paris during the 1780s...Etteilla was the first to produce a bespoke tarot deck specifically designed for occult purposes around 1789. In keeping with the unsubstantiated belief that such cards were derived from the Book of Thoth, Etteilla's tarot contained themes related to ancient Egypt."

Dice, coins, and cards have been used since antiquity for games and divinations, where the fate of the seeker is left to the gods (or nature, or chance, or whatever makes you happy). Lotteries, or drawing "lots", are also activities that rely on random drawings and are used as mechanisms to generate revenue, sometimes to randomly give away land or money, and sometimes as sucker bets.

Digital N/Shutterstock.com

FIGURE 6 - Tarot cards

LOTTERIES

The English word lottery comes from the Dutch noun "lot" meaning "fate". Lottery tickets from the Han dynasty in China between 205 and 187 B.C. are the first known use of lotteries, which presumably financed government projects like the Great Wall of China. The Bible describes casting lots to divide up land. A lottery organized by Roman Emperor Augustus Caesar used revenue for infrastructure repairs in Rome. Lotteries with monetary prizes, such as the King's Lottery, discussed earlier, were held in European towns in the 15th century to raise money for fortifications and to help the poor. A record dated 9 May 1445 at L'Ecluse refers to raising funds to build walls and fortifications. In the 17th century it was common in the Netherlands to organize lotteries to collect money for the poor and in order to raise funds for a wide range of public uses. Then, as now, lotteries were hailed as a painless form of taxation (They're not painless for those who go broke buying losing tickets). The Dutch *Staatsloterij* is the oldest active lottery, running since 1726.

The first Italian lottery was held in Milan in 1449 to help finance the war against the Republic of Venice. *Lotto* became very popular in Genoa, where twice a year people bet on the names of Council members who were then selected by chance, like a random election (Maybe random elections should be tried in the U.S.). When citizens wanted to bet more often than twice a year, the lottery operator substituted candidates' names with numbers. This was the forerunner of modern lotteries and one of the world's earliest sucker bets.

BOTTOM LINE

Cards, coins, dice, lots, and other random number generators, have been around since antiquity and have been used to play games, gamble, and make divinations. In many cases, these devices have been replaced by apps.

CHAPTER 5

A BRIEF HISTORY OF PROBABILITY

"In the long run, the coin will land heads about half the time."
Anonymous gambler.

"In the long run, we are all dead" John Maynard Keynes.

Triff/Shutterstock.com

In 1654, the titled French con man Antoine Gombaud, the Chevalier de Mere, bet that in four rolls of a die, at least one 6 would come up. The Chevalier thought that since the chance equals 1/6 that a 6 would come up when a single die is rolled, then in four rolls, the chance of getting at least one 6 equals 4/6 or 2/3. The Chevalier was wrong (the probability of getting at least one 6 in four rolls equals 0.52), but, in repeated play, he won anyway. When the betting dropped off, the Chevalier introduced a second game. He reasoned that since the chance equals 1/36 that a double-6 will come up when a pair of dice are rolled, the chance that at least one double-6 will come up in 24 rolls equals 24/36 = 2/3. The Chevalier was wrong again, and this time he started losing. In desperation, he consulted the mathematician Blaise Pascal. Pascal's interest in the Chevalier's games spurred correspondence with the mathematician Fermat. This marked the beginning of probability theory as we know it today.

Different Types of Probability

There are different types of probability: Counting combinations, data based, subjective, scientific inquiry, to name a few. In 1933, the famous mathematician, Andrey Kolmogorov, published an abstract, axiomatic definition of a "probability space," which is still used today but which needs some mathematical background to understand.

There is a famous law of probability, called the *law of averages,* or the *law of large numbers*, which states that, given certain conditions, sample fractions and averages stabilize as the sample size gets large.

The Law of Averages

The law of averages, aka the law of large numbers, was introduced by the Italian mathematician Gerolamo Cardano sometime in the 16th century when he observed that the accuracy of averages seems to improve with the number of trials. A proof was later given by Jacob Bernoulli in 1713, who restricted his result to "Bernoulli trials", which are independent trials with just two outcomes, "0 or 1," "heads or tails," "win or lose," etc. A century later, the French mathematician Poisson generalized the result and later the results were further generalized. The basic idea is that in a large number of independent trials of an experiment like coin tossing or making bets in a casino, the *fraction* of times an event occurs gets closer to its probability. For example, if you toss a coin repeatedly, the fraction of heads, or tails, will eventually get arbitrarily close to 0.5. If you toss a coin 100 times, there's about a 95% chance that the fraction of heads will be between 0.4 and 0.6 (40-60 heads).

The law of averages does not mean that in coin tossing the fraction of heads will eventually equal 0.5, but instead that the fraction of heads will get arbitrarily *close* to 0.5, so close, however, that the fraction will, at some point, be effectively indistinguishable from 0.5. You can go out to as many decimal places as you want, and eventually, *rounded off*, the fraction of heads will equal 0.5. Most people would consider 0.4999999999999999 or 0.5000000000000001 to be equal to 0.5 for practical purposes. The law of averages says that eventually the fraction of heads will get even closer to 0.5 than either of these fractions and stay that way.

The historical home of the law of averages is the temple of chance, the casino. Popular gambling games such as craps, roulette, and keno all involve independent trials, and the law of averages can be used to see what will happen "in the long run," or perhaps, in the "medium run," so as not to run afoul of John Maynard Keynes, who said that in the long run we are all dead. In this context, probability is a time-based measure since it gives the fraction of occurrences of an event over time. A very rare event happens only occasionally and so if conditions are met to apply the law of averages, we may not be able to avoid Keynes, and not observe the fraction specified in the law of averages in our lifetime. For example, in the King's Lottery, our lifetime is not long enough to observe the fraction of winning tickets approaching the correct probability, although 0 is a good approximation to the non-zero probability of winning (1 in 3.5 quintillion).

The chance of a major earthquake in San Francisco is very small and, unlike a coin or a pair of dice or the King's lottery, major earthquakes are not necessarily independent events with the same probability of occurrence each time, so this process may not fit into the framework required by the law of averages.

Here are two examples of independent events:

1. Toss a coin twice. The result of the second toss is independent of the result of the first toss.

2. Disturbed by the rise in terrorism, a gambler decides that whenever she travels by plane, she will bring a bomb with her. She thinks that although it is unlikely that there will be a terrorist with a bomb on her plane, it is *very* unlikely that two people will bring bombs on the plane. Unfortunately, what the gambler does is independent of what a terrorist might do, so the probability that a terrorist brings a bomb on the plane is unchanged by the gambler's bizarre strategy.

The Gambler's Fallacy

Suppose you make ten, independent bets at the craps table and lose them all. According to the *gambler's fallacy*, your chance of winning has now increased so that things will average out in the long run. It follows that you should make a big bet in these situations. It's not true. Your chance of winning remains the same. Suppose you toss a coin over and over and get ten heads in a row. You might think tails are "due" because, according to the law of averages, the fraction of heads will eventually be close to 0.5. Although it's true that at some point the fraction of heads will be close to 0.5, the law of averages says nothing about what will happen on the next toss. Regardless of what has happened on previous bets or previous tosses, the probability of winning the next bet or getting heads on the next toss is always the same. The belief that probabilities change with independent events so that things average out causes gamblers to do silly things that they might not ordinarily do, like making big bets when they lose or buying many lottery tickets when they think certain numbers are due.

Dzikovski graphics/Shutterstock.com

Cherry Picking

Another bad habit is *cherry picking*, looking only at wins and not at losses. Even in a game like craps, where the gambler will eventually start losing at a predictable rate, there will be lucky winning streaks along the way just due to chance. The gambler will point to the lucky streaks as evidence of a winning strategy and not mention the offsetting losses. Many a gambler has told a spouse or a friend, "I won $500 today," without mentioning the $2,000 they lost. This happens in the stock market and other areas where there is randomness and where people are taking risks, namely, everywhere. We're all gamblers. Cherry picking is a feel-good thing. Everyone wants to highlight winnings and play down losses, but this can interfere with understanding randomness, in which weird patterns, both winning streaks and losing streaks, occur by chance.

The Chance of Being Struck by Lightning (or Killed by a Bee)

"I never promised you a rose garden, but if you happen to visit one during a thunderstorm, you may be in for a shock." – Mother Nature

You often hear things like: "You're more likely to get struck by lightning than you are to win the lottery." Although this may be true in a sense, the probability of being struck by lightning is a different type of probability than the probability of winning a lottery. The chance of being struck by lightning is usually given

Thomashonnnn/Shutterstock.com

to be about 1 in a million. Where do they get this? Who is "they"? In this case, "they" is the National Weather Service, which keeps track of things like lightning strikes. According to the National Weather Service website, in 2019 the population of the United States was 330 million. Averaging annual weather data over the previous 10 years, from 2009 through 2018, there were an average of 27 deaths and 243 injuries per year for an annual average of 270 people who were struck by lightning. Divide 270 by 330,000,000 to get 1 in 1,222,222, or 1 in 1.22 million, which you can round off to roughly 1 in a million as the current National Weather Service estimate of the probability of being struck by lightning in a year. If you divide 1,222,222 into 302,575,350, it seems that you're 248 times more likely to get struck by lightning than you are to win the Mega Millions Lottery jackpot if you buy one ticket. It's not quite so simple.

Lottery probabilities are easy to understand because every ticket has the same chance of winning and drawings are independent. In a sense, the National Weather Service makes a similar assumption for lightning strikes: that every person in the country has the same chance of being struck by lightning. However, that's not true. In this case, bad luck isn't the same for everyone. Some people are more likely to be struck by lightning than others. If you're standing on a golf course in Florida during a thunderstorm, you're more likely to be struck by lightning than if you're playing roulette in a casino in Nevada. Same for being killed by a hornet, wasp, or bee sting. So, the lightning strike probabilities are just rough averages with assumptions about equal chances that aren't true.

Lightning Strikes at Random, Sort of

On August 4, 2022, a sudden thunderstorm developed late in the afternoon in Washington, D.C. Four people took shelter under a large tree in Lafayette Park, across the street from the White House. Suddenly, lightning struck the tree. As the Washington Post put it,

> "Strangers huddled together under a tree. Just before 7 p.m., it was at that spot — under a leafy tree about 100 feet from a statue of President Andrew Jackson — that lightning struck. The blast killed a retired couple from Wisconsin and a bank official from California. A woman raising money for refugees was badly wounded but survived." (https://www.washingtonpost.com/dc-md-va/2022/08/07/lightning-strike-lafayette-square-victims/) .

Experts recorded a lightning flash in the area as six individual surges of electricity that hit the same point in the space of half a second. If the electricity struck the tree first, experts said, it would have sent hundreds of millions of volts coursing through it before passing into and over the bodies of those gathered beneath it. "It shook the whole area," an eyewitness later recounted. "Literally like a bomb went off, that's how it sounded."

Lightning strikes are brutal events, but they only take place during certain weather conditions and only affect a few, very unlucky people. If you stay inside during a thunderstorm, you won't get struck by lightning, but it's not always that easy. These four people were random victims, brought together by the violent weather that suddenly materializes on late, summer afternoons in Washington, D.C. and that, as far as I know, has nothing to do with politics.

THE CHANCE OF BEING KILLED BY A HORNET, WASP, OR BEE STING

The National Safety Council keeps track of the odds of a wide variety of deaths, including death by drowning, sunstroke, choking on food, sharp objects, hot

Darios/Shutterstock.com

surfaces, dog attacks, falling, and hornet, wasp, or bee stings. The probabilities, or odds as they call them, are calculated the same way as the National Weather Service calculates the probability of being struck by lightning, namely, by dividing the number of occurrences by the population of the U.S. (At the time of this writing the NSC lists the population of the U.S. as 329,484,123, rather than the rounded off 330 million used by the National Weather Service). The NSC website provides death data for 2020 and then displays probabilities as "lifetime odds of death" based on a 77-year lifetime. For example, the estimated probability of being killed by a hornet, wasp, or bee sting sometime in the next year equals 1 in 4,452,525, somewhat less than the estimated probability of being struck by lightning. The NSC realizes that their estimates need an explanation. They say:

> "The odds ... are statistical averages over the whole U.S. population and do not necessarily reflect the chances of death for a particular person from a particular external cause. Any individual's odds of dying from various external causes are affected by the activities in which they participate, where they live and drive, and what kind of work they do, among other factors."

This is good advice. It applies to lightning strike "probabilities" (measures of bad luck) as well as to the probability of getting killed by a hornet, wasp, or bee sting.

Major Earthquakes

Like lightning strikes, major earthquakes are not like casino games or lotteries where we can count combinations in a structured setting. Since major earthquakes are so unlikely, you can't treat them in the same way as lightning strikes, or bee stings either. Some researchers use complex, seismic models to predict these geologic events. Professors D.A. Freedman and Philip Stark from the University of California at Berkeley wrote a paper describing the difficulties of making such predictions. In their introduction, they say,

> "What is the chance that an earthquake of magnitude 6.7 or greater will occur before the year 2030 in the San Francisco Bay Area? The U.S. Geological Survey estimated the chance to be 0.7 ± 0.1. In this

paper, we try to interpret such probabilities. Making sense of earthquake forecasts is surprisingly difficult. In part, this is because the forecasts are based on a complicated mixture of geological maps, rules of thumb, expert opinion, physical models, stochastic models, numerical simulations, as well as geodetic, seismic, and paleoseismic data. Even the concept of probability is hard to define in this context. Despite the work involved in the USGS forecast, their probability estimate is shaky, as is the uncertainty estimate."

Bottom Line

The Chevalier had a 52% chance of winning the first game but only a 49% chance of winning the second game. There is no skill involved in either game, however, in games like this, long run win percentages are predictable and luck eventually runs out. If a proper betting strategy is used, the slight change in probabilities marks the difference between riches and ruin in repeated play. If a proper betting strategy isn't used, you'll go broke no matter how good the game is. For example, if you bet everything you have on each play of a game you'll go broke as soon as you lose one bet.

Probability is how we measure randomness. Under certain conditions, the law of averages brings stability to random processes. The gambler's fallacy and cherry picking illustrate ways that probability can be misinterpreted and distorted. There are many random events outside the comfy confines of a casino, like lightning strikes and earthquakes, where we can't count combinations of equally likely events the way we can with dice and lotteries, and for which finding accurate probabilities becomes much more complicated

CHAPTER 6

CRAPS AND EXPECTED VALUE

"Ever since that big craps game, I been livin' on chicken and wine. I'm a leader of society since I got mine." – Ry Cooder. "Chicken Skin Music"

serpeblu/Shutterstock.com

Probabilities describing casino games, unlike the probability of being struck by lightning, being hit by a truck, or being stung by a bee, which are estimated from data, have a clear interpretation in which the law of averages can be applied. To further understand the law of averages, let's go to the craps table.

Craps

The 17th-century dice game offered by the Chevalier de Mere was one of the origins of modern dice games. In the casino game of craps, a pair of dice are rolled by one of the gamblers and the upturned dots are counted. Before the dice are rolled, you can make a variety of bets. For example, you can bet that the sum of the upturned dots equals 7. If you bet $1 and win, you win $4. If you lose, you lose your dollar. In other words, "payoff odds" for a bet on 7 are 4 to 1. Since there are 6 ways to get 7 (1-6, 2-5, 3-4, 4-3, 5-2, 6-1) and 36 outcomes altogether (6 faces on a die, so 6x6 =36 outcomes), your chance of winning a bet on 7 equals 6/36 = 1/6, and the chance of losing = 5/6. There is no skill involved. It's all luck. You can practice 24 hours per day and a chimpanzee will do just as well as you, better if it gets lucky.

In this game, a gambler's fortune hinges on luck, until, after enough bets, luck, like time, slips away. Since dice rolls are independent, with each roll unaffected by previous rolls, there is no way, other than possibly by cheating, to do well in this game in the long run (You can buy trick dice on Amazon for about $10, but it's not a good idea to use them in a casino). According to the law of averages, if you play repeatedly, the fraction of times you win bets on 7 will get closer and closer to 1/6, the chance of 7 coming up on one roll, and the fraction of times you lose will get closer to 5/6. It follows that in repeated $1 bets on 7, you will win $4 (payoff odds) about 1/6th of the time and lose $1 about 5/6th of the time, for an average, or "expected value," or EV, of 4/6 – 5/6 = -1/6, or about -$.17, a 17 cent loss per dollar bet. In the grand tradition of sucker bets.

Expected Value

This bet has negative "expected value" so you will eventually lose, regardless of the strategy you use. On the other hand, a bet with positive expected value allows the gambler to make a long-run profit, but there aren't many bets in the casino in which a gambler has a positive expected value. None at the craps table. And if you happen to find a game with positive expected value, you can still lose if you don't have a good betting strategy.

In the case of most casino games, the negative of the player's expected value is the casino's expected value and is called the "house edge." The house edge is the *casino's* long-run average profit. Craps is a "zero-sum" game: The player's losses are the casino's wins, so the casino has a positive expected value. As far as the casino is concerned, "long run" could mean one player playing many times or many players playing a few times. As long as there is plenty of action, the house edge is a reliable indicator of the casino's profit. For games like craps, the chance of a player ending up a winner by getting lucky gets smaller as the number of bets gets larger. This is because of the law of averages, which guarantees that after a while every gambler who isn't cheating gets close to *average*, which, in this case, means losing 17 cents per dollar bet. In this context, luck is a short-term concept. Eventually, the average result is predictable. There's nothing you can do about it except find something else to do.

What about bad luck? In a game like craps, you can think of bad luck as losing more than average, namely, doing *worse* than an average loss of 17 cents per dollar bet (as if that's not bad enough). But every cloud has a silver lining. You might do worse than average for a while, but eventually, the law of averages guarantees that your average losses will claw back up to about 17 cents per dollar bet. Maybe you should count your losses as entertainment expenses rather than misfortune. Or the cost of doing business. Maybe you should do that about everything. If there are a large number of gamblers playing craps, then, just like lottery winners, there is the opportunity for an occasional gambler to get lucky and be a winner. But even the occasional lucky winner will eventually start losing if they keep playing.

Expected Value of the Mega Millions Lottery

Lotteries like the Mega Millions Lottery have a payoff structure different from casino games like craps. Instead of winning or losing every time there is a lottery drawing, the Lottery Commission simply takes 50% of the revenue from ticket sales, with the remaining 50% being used to pay off winners. The Lottery Commission uses the money it takes to pay administrative costs, advertising costs, equipment costs, and other costs incurred in the course of running the game. The Commission also gives each state a percentage of the revenue, which is used for whatever purposes the state wants. Most of the states use at least some of this money to support education. In any event, the house edge for the Mega Millions lottery is 50%. It's not just an average, it happens with every drawing.

The Inner Oracle

A successful car salesman in Minnesota is on his way to work when he hears a voice in his head: "Quit your job." The man looks around and doesn't see anything unusual, so he proceeds on his way to work. Then he hears the voice again: "Quit your job." The man ignores the voice and continues to work, but the voice persists. He can't get rid of the voice. Finally, after a week of being driven crazy by the voice, the man submits his resignation.

Nothing happens for a few days, but then the voice returns: "Sell your house." The man is beside himself. He doesn't know what to do. He goes to the doctor, but the doctor attributes the voice to his imagination. When he gets home, the voice returns: "Sell your house." This goes on for a week. Finally, the guy can't stand it anymore, so he sells his house and rents a room in a motel. The voice returns. "Close out all your bank accounts and put the money, along with the money from selling your house, along with your severance pay, into cash." The man does as he is told. He's sitting in his motel room with a briefcase filled with cash when the voice returns: "Take your cash, go to the airport and take the next flight to Las Vegas."

At this point the man isn't going to argue with the voice, so he goes to the airport and takes the next flight to Las Vegas. When his flight lands in Las Vegas, the voice instructs the man to take a taxi to a large casino on the Las Vegas strip. The man does this. When he enters the casino, the voice tells him to go to the craps table and bet all his money on #7. The man protests but the voice is persistent and yells at the man to do as he's told. Finally, the man walks over to the craps table and bets all his cash, the proceeds from his house, his severance pay, every penny he has, on #7. The dice are rolled. #11 comes up.

There is a moment of silence as the man stares at the dice in disbelief. Then the voice in his head says, "Oh s**t!"

Bottom Line

Craps is a game of chance. There is no skill involved. The only way you can win is by luck. Expected Value gives a gambler's average winnings or losses per bet for a particular gamble in repeated, independent plays. According to the law of averages, for games like craps, luck eventually disappears. Since the expected value of a craps bet is negative, a persistent craps bettor will eventually go broke.

CHAPTER 7

ROULETTE - INCLUDING THE DOUBLE-UP AND KARA-OKE STRATEGIES

"Feeling as though I were delirious with fever, I moved the whole pile of money to the red – and suddenly came to my senses! For the only time in the course of the whole evening, fear laid its icy finger on me and my arms and legs began to shake. With horror I saw and for an instant fully realized what it would mean to me to lose now! My whole life depended on that stake!" …

I elbowed my way into the thick of them and stood close to the croupier;… It appeared to me that pure calculation means fairly little and has none of the importance many gamblers attach to it. They sit over bits of paper ruled into columns, note down the coups, count up, compute probabilities, do sums, finally put down their stakes and – lose exactly the same as we poor mortals playing without calculation." – "The Gambler" - Fyodor Dostoyevsky,

Studio Romantic/Shutterstock.com

"Making a combination of bets with negative expected value is like the retailer who sells every item at a loss, hoping to make up for it in volume."
– Anonymous gambler.

Roulette Basics

An American roulette wheel is divided into 38 sections, numbered from 1 to 36, 0 and 00. 18 of the sections numbered from 1 to 36 are black and 18 are red. Sections 0 and 00 are green. A European wheel doesn't have a section marked 00, which is good, but not good enough for positive expected value. Unless you're Dr. Richard Jarecki, whom I'll discuss in a minute. The wheel is spun by a casino employee, who then spins a ball along the wheel in the opposite direction. As the wheel and the ball slow down, the ball drops into the winning section on the wheel, and bets are decided.

You can bet on individual numbers, colors, and any combination of numbers and colors. For example, you can:

Bet on #17, P(win) = 1/38 (1 section in 38), P(lose) = 37/38

Bet on red, P(win) = 18/38 (18 sections in 38), P(lose) = 20/38

You might think that deciding which of these two bets to make is a no-brainer since the chance of red coming up is 18 times greater than the chance of #17 coming up. However, when making bets in a casino and elsewhere, you must consider both the chance of winning and the payoff odds. In this case,

Payoff odds for bets on #17 are 35 to 1

Payoff odds for bets on red are 1 to 1, or "even money"

You can compute the expected value for a $1 bet using the same method I used for craps. If you repeatedly bet $1on red, the law of averages says that red will come up an average of 18 times in 38 spins. So, on average, you'll win about 18/38 of your bets in repeated play.

expected value = $1 * (18/38) - $1 * (20/38) = - $2/38 = - $.053

If you bet on red repeatedly, although there is random variation from bet to bet, in the long run you will lose at an average rate of 5.3 cents per dollar bet. Roulette is a sucker bet, unless the entertainment value is worth more to you than 5.3 cents per dollar bet.

Suppose you bet $1 on #17. The law of averages says that in repeated play #17 will come up about once in 38 spins, so on average, in 38 spins, you will win 1 bet and lose 37. The payoff odds are 35 to 1, so:

$$\text{expected value} = \$35 * (1/38) - \$1 * (37/38) = -\$.053$$

In repeated play, your average loss will be 5.3 cents per dollar bet, the same as bets on red! Both bets have the same negative expected value. If you must make one of these bets, your choice of which bet to make might be based on personal preference. For example, if you like to win occasionally and have a big payoff when you do win, betting on #17 might be your preferred bet, whereas if you like to win frequently and don't care about a big payoff, betting on red might be the preferred bet for you. Don't forget that whichever of these bets you like, you will eventually lose.

Goritza/Shutterstock.com

It turns out that no matter what strategy you use in roulette you will eventually lose at the rate of 5.3 cents per dollar bet (There is an exception, called the five-number bet, but it's worse). Given enough betting action, the casino will make a profit of about 5.3% of all money bet.

The Double-Up Strategy (Sometimes Called the Double-Down Strategy)

A popular and deceptive roulette strategy that is also popular outside the temple of chance (casino) is the "double-up" strategy, which goes as follows:

- 1st bet – Bet $1 on red. If red comes up, you win $1. Quit.
- 2nd bet (if necessary) – If red *doesn't* come up, double your bet and bet $2 on red. If red comes up, you win $2, covering your $1 loss on the first bet and leaving you with a $1 profit. Quit.
- 3rd bet – If you lose the second bet as well as the first bet, double your bet and bet $4 on red. If red comes up, you win $4, covering your previous $1 and $2 losses and leaving you with a $1 profit. Quit.
- Etc.

In other words, you keep betting until red comes up, doubling your previous bet each time you lose. According to the laws of probability, red eventually will come up, usually (but not always) in the first few spins. Doubling your bet after each loss cancels out all previous losses and guarantees you a net win of $1 when red finally comes up, at which point you quit. Is this a guaranteed winning strategy?

All casino games have a betting limit. If you encounter an unlucky streak of losses, the amount you need to bet may exceed this limit, causing you to not be able to cover your losses. In addition, most gamblers have a personal limit. If you encounter an unlucky streak of losses, the amount you need to bet may exceed this limit, also causing you not to be able to cover your losses. This can set the stage for a big loss. For example, if you're really unlucky and red fails to come up 15 times in a row, on the 16th bet, you must wager $32,768 in an attempt to come out $1 ahead. The double-up strategy is deceptive because most of the time red will come up within the first few bets, and you will win $1 (or whatever initial bet you started with) and hopefully quit. Unfortunately, on the rare occasions when you can't cover your losses, you will incur a big loss that will wipe out your previous, small wins.

Versions of the double-up strategy are used in a variety of situations. It's not uncommon to see sports bettors, stock market investors, and degenerate gamblers of all types bet larger and larger amounts to cover their losses. It turns out that even in games with positive expected value, it's not a good idea to make bigger and bigger bets to cover your losses.

The Karaoke Strategy

The double-up strategy says to *quit* as soon as you win. What does it mean to quit? Does it mean that as soon as you win you will never come back to the roulette table? Or does it mean you will go to the karaoke bar for a break and then start over? This is called the *karaoke strategy*. The karaoke strategy is fine except for the "start over" part. In games like craps and roulette, you're better off if you stay at the karaoke bar for the rest of your life.

Watch the Wheel

Stand around the roulette table until you see 10 reds in a row. Then, what should you do?

1. Bet on black, because "black is due."
2. Bet on red, because "red is hot."
3. It's all random. Don't bet.

Option 1) is an example of the gambler's fallacy, mentioned earlier. In repeated, independent trials of a random experiment, like repeatedly spinning a roulette wheel, nothing is due. The law of averages specifies what will happen in the long run, not on an individual spin of the wheel.

Option 3) is a good choice. Randomness has streaks. If you see 10 reds in a row, it's probably a random thing, especially if you've been watching the wheel for a long time.

Option 2) makes sense on rare occasions. If there is something wrong with the equipment, you might be able to detect a departure from randomness by watching the wheel and analyzing the observed data. The chance of seeing 10 reds in a row is small (1 in 1,758). If you watch the wheel long enough, you might be able to decide if this (or some other pattern) has happened due to chance or due to a defect in the wheel. A statistician's challenge is to determine if something is due to chance or if observations are so unlikely under the chance hypothesis that you would reject chance in favor of a plausible alternative. If you decide to reject randomness and conclude that the equipment is faulty, based on how unlikely it is to have red come up 10 times in a row, it might make sense to bet on red. Probably not a good idea in a casino with well-maintained equipment, but this strategy paid off for Dr. Richard Jarecki, a physician from New Jersey who made serious money betting on roulette in the 1960s and 1970s in European casinos where there is no 00 and where the casino advantage is 2.7% compared to the 5.3% in American casinos.

Dr. Jarecki

Dr. Jarecki made a small fortune by "watching the wheel," keeping track of the results, and betting at roulette tables where the results were not quite random due to wear and tear or other mechanical defects. According to a New York Times article:

> "Ms. Jarecki said in a telephone interview on Monday that she, Dr. Jarecki and a handful of other people helping them would record the results of every turn of a given roulette wheel to discover its biases, or tendency to land on some numbers more frequently than others, usually because of a minute mechanical defect caused by shoddy manufacturing or wear and tear."

The article further goes on to say,

> "Watching, or 'clocking,' a wheel ...could mean observing more than 10,000 spins over as long as a month. Sometimes a wheel would yield no observable advantage. But when Dr. Jarecki and company did find a wheel with a discernible bias, he would have an edge over the house."

As a result of Dr. Jarecki's success, casinos in both Europe and the U.S. became more careful about the maintenance, quality, and replacement of their roulette wheels. Today, the kinds of opportunities Dr. Jarecki discovered are not readily available, but who knows?

THE ADDITION RULE - YOU CAN'T TURN A GROUP OF BAD BETS INTO A GOOD BET

I was watching the roulette table at a casino in Las Vegas, and I saw someone put a chip on every number in the betting layout: 0, 00, 1, 2, 3, ...all the way through 36. This guaranteed that one of the winning numbers would be selected, guaranteeing a win of 35 times the amount on the chip. As I recall, they were $5 chips, so he won $35*5 = $175. Unfortunately, all the other chips were losses, for a combined loss of $5*37 = $185. The player's net loss was $175 - $185 = -$10 with a total amount bet equal to $5*38 = $190. Divide -$10 by $190 to get -$0.053, the player's average loss per dollar bet. You probably noticed that this is the same negative expected value a player has for making (almost) any roulette bet. In this case, however, the gambler would lose that amount on every spin of the wheel. Randomness isn't even a factor in determining this gambler's results!

More generally, a gambler's overall expected value for making *any* combination of roulette bets, or any other bets is just the sum of the expected values of the individual bets. If all the individual bets have negative expected value, then so does the combination of these bets, which is just the sum of the individual expected values.

You can't make a positive by adding a bunch of negatives. It's like the retailer who sells every item at a loss, hoping to make up for it in volume.

Here's a further explanation. Suppose someone bets $1 that "7" will come up in craps (4 to 1 payoff odds) and $1 that red will come up in Roulette (1 to 1 payoff odds). What's the expected value of the combined bet? At first glance, it seems that you need to look at all possible combinations that result from making these two bets, find the probability of each, and then compute the expected value of the combined bet. You would start with this table and then do the necessary computations to get expected value.

outcome	probability	payoff
win craps, lose roulette	(1/6)*(20/38)	4-1=3
lose craps, win roulette	(5/6)*(18/38)	1-1 = 0
win craps, win roulette	(1/6)*(18/38)	4+1 = 5
lose craps, lose roulette	(5/6)*(20/38)	-2

And that's just for two bets. Fortunately, you don't need to do any of this. The "addition rule" for expected value says that if you have a situation like this, all you need to do is to add the individual expected values. We already know that the expected value for a $1 bet on "7" in craps is $-.17 and the expected value for a $1 bet on red in roulette is $-.053. It follows that the combined bet of $1 on "7" and $1 on Red has expected value equal to $-.17 + $-.053 = $-.223 for a $2 total bet, or $-.223/2 = $-.1115 per dollar bet. If you made numerous bets on numerous things, you could do the same thing.

Gamblers, investors and others like to put together complex strategies of combination bets, trying to put a group of bad bets with negative expected value into a good bet with positive expected value. (This is not the same as a successful hedge fund, which is a combination of bets mostly with positive expected values). If you've ever observed the betting at the roulette table, you will have seen gamblers making many bets at once, presumably hoping to win some bets when others lose. The addition rule for expected value says that on the average, the net loss per dollar bet equals $-.053.

Pie Throwing

Ljupco Smokovski//Shutterstock.com

There are 20 clown school students seated in a circle, all of whom have pies. Each clown tosses a coin. If the coin lands heads, the clown hits the person to their left with the pie. If the coin lands tails, the clown hits the person to their right with the pie. What's the expected number of clowns who don't get hit by a pie? The addition rule for expected value saves the day.

Ellen is one of the clown school students. She doesn't get hit by a pie if both the person on her left and the person on her right throw their pies at the person away from Ellen. Since this is based on a coin toss, the probability Ellen doesn't get hit by a pie equals 0.5*0.5 = 0.25. The "indicator" for Ellen is 1 if she doesn't get hit by a pie and 0 if she does. The expected value of an indicator function is 1*P(it happens) + 0*P(it doesn't happen) = P(it happens), which in this case is the probability that Ellen *doesn't* get hit by a pie, namely, 0.25. This is true for everybody in the circle. To find the expected number of clowns who don't get hit by a pie, use the addition rule for expected value, which becomes 20*.25 = 5. In other words, we expect about 5 clowns to not get hit by pies. In general, if there are n people, and if P(indicator = 1) = p, then expected number of indicated events equals n*p.

Chickens in a Circle

There are 30 chickens sitting in a circle. Each chicken randomly pecks the chicken to its right or the chicken to its left. On average, how many chickens don't get pecked? What if there were 50 chickens? Same method as clowns with pies, in this case, 30*0.25 = 7.5, and 50*0.25 = 12.5.

Mailboxes

There are 5 mailboxes. You have 6 letters to mail. Each time you mail a letter, you pick one of the mailboxes at random. You mail all the letters this way. What is the expected number of mailboxes left empty? Since P(a letter is not mailed in a particular mailbox) = 4/5 = 0.8, it follows that P(particular mailbox is left empty) = 0.8^6 = 0.262. Using the addition rule for expected value, the expected number of mailboxes left empty equals 0.262*6 = 1.57.

Bottom Line

Roulette is a game of chance. There's no skill involved. If the wheel is properly balanced and nobody is cheating, the only way you can win is by being lucky, and even luck disappears eventually. Combining many bets that all have negative expected value results in a combined bet whose expected value is the sum of the individual bets and so also has negative expected value.

CHAPTER 8

BLACKJACK, POSITIVE EXPECTED VALUE, AND THE KELLY SYSTEM

"To make money, buy some good stock, hold it until it goes up, and then sell it. If it doesn't go up, don't buy it." – Will Rogers.

Blackjack, or "21," is a popular casino card game in which players play against a dealer to get a card total as close to 21 as possible without going over 21. Although blackjack, or some version of it, has been around for a long time, strategies that gave the player a positive expected value were first developed by researchers at IBM by doing computer simulations in

juliannedev/Shutterstock.com

the late 1950s. These strategies were publicized by math professor Ed Thorp, whose 1962 book, *Beat the Dealer*, still in print, caused thousands of gamblers to flock to the casinos and casino owners to flock to the pharmacies to get antidepressants.

Fortunately for the casinos, winning strategies, which involved "card counting," keeping track of the cards as they were dealt from the deck, were difficult and gave the player only a slight edge under ideal conditions. Casino countermeasures were introduced, such as multiple-deck games and frequent shuffling, Most blackjack players lost. Pros were banned from the tables. Sophisticated players responded with strategies to retain their edge. It's not clear that they were doing anything illegal, but casinos reserve the right to throw out undesirables (It would be undesirable from the casino's point of view if players were consistently winning.)

Computer Simulations

Computer simulations were first developed by Stanislaus Ulam, a mathematician who worked on the development of thermonuclear weapons at the end of World War II and who later used computers, still in their early stages of development, to devise the "Monte-Carlo" method, named after the casinos of Monte Carlo. Simulation software typically uses a computer's random number generator and lets you approximate probabilities by repeating a process a large number of times. Simulated probabilities can then be used to estimate theoretical probabilities and to solve problems too complicated for traditional methods. With today's fast computers, simulation has become an important tool in the statistician's toolbox for approximating solutions to complex problems, making predictions, and addressing other statistical challenges.

How is Roulette Different from Blackjack?

Unlike roulette and craps, which involve independent spins of a roulette wheel or rolls of a pair of dice, blackjack has "conditional probabilities", or dependent events, depending on which cards have been dealt from the deck (more on conditional probabilities later). It's difficult to use the law of averages in games like this, because as cards are dealt, you don't have a sequence of independent events. It gets complicated when a blackjack player has to decide what to do for any given configuration of the deck because there are so many configurations. *Simulations* allow many independent trials of a strategy for a given configuration of the deck, thus providing approximate probabilities for winning or losing using a particular strategy and starting with that configuration. By using simulation software, Thorp and others were able to play millions of simulated blackjack hands for different configurations of a deck and compare various strategies to determine which strategies were best (the dealer plays by fixed rules). For example, should you "stand" or "draw" when you have 10 and 6 and the dealer is showing a Queen? A simulation quickly plays millions of blackjack hands for each choice (stand or draw) to see which choice is better. The blackjack simulations of the 1950s produced strategies that require memorizing types of cards as they're dealt from a deck and provide the player with a game that, unlike other casino games, has positive expected value. No wonder the casinos modified the rules a bit.

The Kelly Betting System

If you use a winning blackjack strategy, you will have roughly a 2% positive expected value under ideal playing conditions (long run average winnings equals 2 cents per dollar bet). There are some configurations of the deck that favor the casino and some that favor the player, however, so a skilled blackjack player must be patient and prepare to play for long periods, keeping track of the cards, waiting for positive configurations, and betting accordingly. Even if you're playing a game with positive expected value, you must be careful how you bet. Generally, you should use a "fixed fraction" betting strategy of the type first developed by J.L. Kelly, a mathematician who worked for Bell Labs in the 1950s. The Kelly system, as it is popularly known, optimizes the rate of growth of the player's bankroll for games with positive expected value. Fixed fraction betting means you always bet a fixed fraction of the amount you have available for betting. The more you have, the more you bet. The less you have, the less you bet. This is the opposite of the double-up strategy, where you bet more when you lose to cover your losses, always a bad idea.

For simple situations where you know or can estimate the probabilities of winning, p, and losing, q, and with payoff odds, b to 1, the Kelly system says to bet the fraction of your bankroll equal to p – q/b. (In blackjack, since b = 1 for most situations, you would bet the fraction p – q). If your chance of losing is greater than your chance of winning, you should bet 0, or a very low bet if you want to keep your place at the blackjack table. If your chance of winning equals 51% and your chance of losing equals 49%, you should bet 2% of your bankroll. If you have $1,000, bet $20. If you have $2,000, bet $40, and so on. Some stock market investors use variations of the Kelly system, but the stock market is more complicated than blackjack. For example, it's difficult to estimate the probability of "winning" an investment.

The Micky Mouse Watch

This story illustrates the length to which gamblers will go and the length casinos will go to stop them when they have a game with positive expected value. One of the problems with successfully using a card counting blackjack strategy with positive expected value is that if the casino catches you card counting, they'll often spoil your fun and ask you to leave. Over the years, card counters have tried various methods to avoid detection. One bettor I know who had been thrown out of every major casino in Nevada decided to change his appearance. Dan, a pasty-faced gambler who went outdoors only to go from one casino to another (may he rest in peace), dyed his hair, went to a tanning salon, and grew a beard. His brother didn't recognize him. He went to a casino and sat down at a table. Five minutes later, the pit boss approached him and said, "Hi Dan. Could I interest you in taking a walk out the front door?" When Dan asked the pit boss how he recognized him, the pit boss pointed to Dan's watch. He had forgotten to remove the Micky Mouse watch his girlfriend had given him a few years earlier.

One method used by casinos to detect card counters is to watch betting patterns. An average bettor won't make a sequence of $5 bets followed by a few $500 bets, then back to $5 bets. (The $5 bets are minimum bets a card counter uses to stay in the game until the deck is "favorable" again.) One gambler I know was thrown out of a casino merely for varying his bets. Ironically, he was losing $1,000 when he was kicked out.

In the 1970's, Frank S., a well-known blackjack player who recently passed away (may he rest in peace), developed a system known as the "big player concept" in order to stop the casinos from noticing bet variations. Using the pseudonym Al Francesco, Frank recruited teams of blackjack players who were all experienced card counters. The initial teams had 7 blackjack players, six card counters and a "Big Player." The team would enter a casino one by one, with the card counters sitting at separate blackjack tables, keeping track of the cards as they were dealt and making minimum bets. Meanwhile, the Big Player would swagger through the casino like a drunken high roller. When the count at a table got high, indicating a good bet, the card counter at this table would signal the Big Player, who would lurch over and make big bets. If the count got low, the counter would signal again, and the Big Player would leave the table. The counters were always making small bets and the Big Player was always making big bets. There were no bet variations that the casino could use to identify a card counter.

Frank used this system for many years, taking teams of blackjack players to casinos around the world. Others copied the Big Player system, including a team of players from MIT, about whom the successful 2008 movie, "21", starring Kevin Spacey, Jim Sturgess, Kate Bosworth, and Lawrence Fishburne, was made. The casinos eventually figured out the Big Player system and put a stop to it. In addition to being known as the "godfather of blackjack," Frank was an avid horse player who watched the races at Golden Gate Fields in Albany, CA, from the balcony of his condo across the freeway from the track.

I knew Frank back in the day through one of my friends, who was the family attorney. Frank was a self-taught genius and an evangelist for blackjack, even when he stopped using the Big Player system. In later years, he tried to market a card-counting computer that would be strapped around the gambler's thigh. Small sensing devices were placed under the bettor's toes for tapping information into the computer as cards were dealt, and a buzzer signaled the bettor to draw, stand, etc. The device, along with a similar device hidden in a pair of sunglasses, worked fine in the lab, i.e., Frank's living room, but was difficult to use in a casino and was never as successful as the Big Player.

The Chinese Restaurant

Around the time that card counters were a threat to casino profits, one of the casinos on the Las Vegas strip was trying to attract more customers. The casino owner decided that a moderately-priced Chinese restaurant would do the trick. The owner and his assistant decided to convert a conference room on the main floor of the casino to a Chinese restaurant. They had almost finished the remodeling process and had applied for a permit for a new kitchen in a separate area behind the dining area. The city turned down the permit, saying that the casino already had the maximum number of kitchens for their other restaurants on the main floor of the casino. The casino owner was frustrated and upset because they had almost finished remodeling the conference room. After consultation with his assistant, the casino owner decided to open the restaurant anyway.

A few months later, after much publicity and fanfare, the new Chinese restaurant opened. The restaurant was packed. A team of waiters went around taking orders from the extensive menu. Upon taking an order, a waiter would go into the back room, presumably to give the order to the kitchen crew. The problem was that there wasn't a kitchen crew. There wasn't even a kitchen. The back room was a staging area. Instead of giving the kitchen crew an order, the waiter would call the Chinese restaurant down the street and give them the order as take-out. Every ten minutes or so, a waiter would leave the casino through a back door, walk down the street, pick up take-out orders from the actual Chinese restaurant, where they got a special price because of the volume of orders, and return to the back room in the casino. There, the waiters would plate the orders and take them to the dining area where they were served as if they were prepared on the premises. The restaurant got a good writeup in the local paper. The operation took the concept of take-out food to a new level.

Bottom Line

Chance is a factor when dealing cards from a deck, however, computer simulations have shown that there can be skill in playing blackjack. Winning strategies involve keeping track of the composition of cards remaining in the deck and playing accordingly. Even if you are a skilled blackjack player, you must also use a betting strategy like the Kelly system in order to insure a long run profit.

CHAPTER 9

RANDOM SAMPLING AND THE LACK OF IT

One of the first random samples was taken by ancient oracles who tossed the yin-yang coins to select a random divination from the I Ching. Due in part to advances in technology and communication, I Ching divinations have been partially replaced by calling people at random and getting information from them. Yesterday's oracles are today's statisticians. In today's world we may be interested in different topics, but the methods are similar. Getting a prophesy and trying to learn where you fit in with the cosmos has been replaced by trying to estimate the percentage of college students who drink cola lite.

New Africa/Shutterstock.com

To be useful, a sample must be representative of the population being studied. The best way to do this is with a random sample. An element of a population is selected at random if every element of the population has the same chance of being selected. A random sample is an extension of this.

This simple definition eliminates bias (mostly) because every element of the population is equally likely to be selected. If the population size is small, it's easy to take a random sample. To randomly select 3 attendees from a birthday party of 32 people, put everyone's name on pieces of paper, put them in a box, mix up the pieces, close your eyes, and pick 3 pieces of paper from the box. You could use a coin, even an I Ching coin, to do this. There are 2^5 = 32 different heads-tails (or yin-yang) combinations when you toss a coin 5 times (HHTHT, THTHT, TTTHH, etc.). Assign a heads-tails combination to each person. Then, toss the coin 5 times and select the person who was assigned to the combination that came up. A coin, I Ching or otherwise, is a random number generator, just like the random number generator in Excel, Google Sheets, or other software. For example, in Excel you can pick a number at random between the numbers a and b by using the Excel function "=randbetween(a, b)". Speaking of software, if all the elements of the population are listed in a spreadsheet, it's easy to take a random sample: Just use the random number generator in the spreadsheet software to randomly select a sample from the listed population. Using the I Ching coins to get a divination or a sample has been around for thousands of years in one form or another. Using software is the modern method.

Even with a group of 32 people, you need to be careful how you take a random sample. Close your eyes and think of a random number between 1 and 32. You probably picked a number in the middle: 10 through 20, for example. When asked to do this, people seldom pick the extremes. Leave yourself out of this. You can't select a random sample by guessing, so why try? You need a system that can be used by others.

Public Opinion Polls

Suppose you want to take a survey of 1,000 randomly selected registered voters in California. A logical way to contact people who live across the state (or even nationwide) is by telephone. In the past, phone surveys were unre-

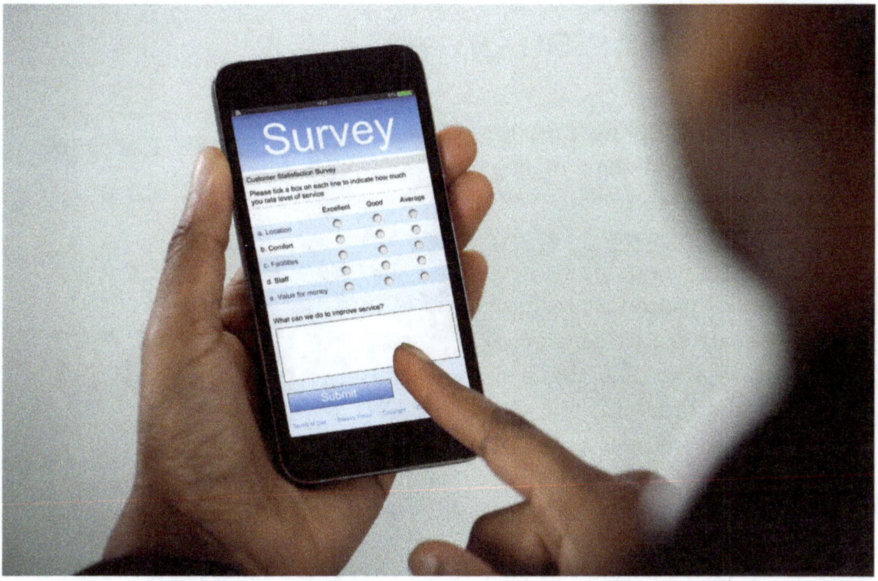

Andrey_Popov/Shutterstock.com

liable because not everyone had a phone. This left out part of the population and so wasn't representative. That's not true anymore. Almost everyone has a smartphone. Sometimes I think that the only person left in the U.S. without a smartphone is Ed, who doesn't like to go along with the crowd. Since most phones today are smartphones and not listed in a phone book, you can't get a random sample by randomly picking phone numbers from a phone book, if such a book even exists.

Random digit dialing has taken the place of the phone book. You select area codes and prefixes for areas you will be calling and dial the last four digits at random. With random digit dialing, in addition to legitimate calls, you will get disconnected numbers, commercial numbers, children, cults, etc. No problem. Just politely hang up (is that an oxymoron?) and keep dialing until you have a random sample of registered voters. This is such a popular method of taking surveys (you never have to leave your home or office) that you can buy phone equipment that will do the random digit dialing for you and even use an automated script when someone answers the phone. For random digit dialing across a large area, like California, it may be necessary to break up the area into smaller, representative parts, sample from the smaller areas and then combine the results like a representative jigsaw puzzle. Doing this properly takes expertise and a research department that can keep up-to-date representative areas. Soon, AI will do this and there will be no need for us.

Face to Face Surveys

Daniel M Ernst/Shutterstock.com

Some types of sampling can't be done using random digit dialing because you need to interact with respondents in person. For example, if you're researching a new cola and want to administer a taste test to estimate the percentage of cola drinkers who like the new cola better than the old cola, you need a face-to-face interaction. A convenient venue for face to face surveys, at least when there's no pandemic, is someplace people congregate, like a shopping mall. This is particularly useful when you are testing products that people buy at the mall because your sample is more likely to be representative than if the product you're testing can only be purchased someplace else. Since some people in the population of cola drinkers don't go to the mall, your mall sample isn't exactly a random sample, but it may provide a representative cross-section of mall shoppers. Because it's convenient, this type of sample is called a "convenience" sample. This will all change in the future, when robot interviewers will interview robot shoppers.

Social Media Polls

Social media is a popular place for polls. They reach a large audience, and the results are ready to process. There's only one problem: They're useless. The reason is that they're not random samples. Instead, they're "self-selecting," in the sense that users of the site decide whether or not they want to participate. This is different from a random sample since not everyone has the same chance of being selected. Typically, only people interested in the topic participate. Passionate people tell their friends to participate. If possible, some people respond multiple times. It's the survey version of the old political adage: "Respond early and respond often."

Bakhtiar Zein/Shutterstock.com

Non-Response

One problem with all surveys, even when run by competent pollsters, is the "non-response" problem, or in a phone survey, the "hang up" problem (As in hang up the phone, not that the respondent has a psychological hang-up. Although maybe they do have a psychological hangup. Why else would they not want to participate in your survey?) Professional surveyors are trained to keep respondents in the survey. If it's a phone survey, surveyors might call back and try to be more persuasive if the respondent hung up the first time. Sometimes, a reward is offered if people participate (gift card, etc.). Or maybe a threat (If you don't respond, you will be inundated by thousands of surveys.) Additional respondents are randomly selected to account for non-response.

Sample Size Matters – Margin of Error

How large a sample do you need to get a good random sample? Depends on what you mean by "good." Suppose you're conducting a public opinion poll in which you ask each respondent, "Do you think we should build colonies on Mars?" Since the results of the sample are used to estimate the percentage of the *population* who think we should build colonies on Mars, hopefully the percentage in the sample will be close to the unknown population percentage. What does "close to" mean? Let's say that you want the sample percentage to be within 3 percentage points of the unknown population percentage. This is called the "margin of error." For example, if the sample percentage of people who think we should build colonies on Mars is 35%, you want to be pretty sure that the corresponding population percentage is between 32% and 38%. The margin of error indicates how much room for random variation you should allocate to be pretty sure that the sample result is accurate, plus or minus that amount.

Siberian Art/Shutterstock.com

What do you mean by "pretty sure?" Suppose that "pretty sure" means 95% confident: In repeated surveys, you want the sample percentage (or proportion) to be within three percentage points of the population percentage about 95% of the time.

We can use our friend, math, to determine how large a random sample we need to be 95% confident that the sample percentage is within three percentage points of the population percentage. It turns out that if the sample size is 1,500 or so, these conditions will be met. It also turns out that for large populations, the sample size needed to obtain a specified degree of confidence and accuracy does not depend on the population size. Think about it. For a given sample size, you get the same amount of information, whether the population size is 1 million, 50 million, or 500 million.

There's a tradeoff between confidence and precision. For a fixed sample size, the more confident you want to be, the wider your margin of error will be, and the higher you want your precision (margin of error), the less confident you will be. You can control both by increasing the sample size.

For large samples, margin of error can be found using the fact that the sample proportion (or sample average) has approximately a normal distribution and that for normally distributed data, roughly 95% of the values are within

2 "standard errors" of the mean. The margin of error for 95% confidence can then be found by multiplying the standard error by 2. More on the normal distribution can be found in Chapter 24.

Election Polls

Election polls are notoriously inaccurate, particularly when there is time before an election. There are three main reasons for this: 1) people change their minds, 2) people refuse to answer, and 3) things change, even in the last week or two before an election. Nate Silver, founder of the website 538.com, who now writes a newsletter on Substack and appears on TV, publishes sophisticated election models that account for background variables and average many polls to make predictions based on these averages. He also does simulations using state-by-state poll averages to get an estimate of the electoral college votes and a prediction of who will win the presidency. These sophisticated models have been more accurate than individual polls with no simulations.

Inspection Bias

Sometimes, just by the nature of the data, you might take a random sample and still over-represent categories. For example, suppose you want to estimate the average class size of Basic Statistics classes at a certain college. You take a random sample of 50 students who are taking Basic Statistics this semester and ask each of them how large their class is. You obtain an average class size of 38. You then look at an Admissions and Records spreadsheet and find that the average class size for Basic Statistics equals 25. The difference is too large to attribute to random variation. There's nothing wrong with your random sample. Or is there?

You are taking a random sample, but even though everyone has the same chance of being selected, more students in your sample will be in large classes just by the fact that they're large classes. When you take a sample like this, even though it's random, you will oversample the large classes and your average class size will be inflated. This is called the *Inspection Paradox*, because your sample is random but by the very nature of what you're inspecting, you get an "oversampled" result. In this case, since you are estimating class sizes, you should be sampling classes not students.

Survivorship Bias

A tourist visiting Copenhagen was walking around town and noticed that everyone looked healthy. Not a sick person to be seen. How could this be true? This is an example of "survivorship bias," a situation in which a presumed random sample isn't really random because the sample elements have "survived" the selection process (being on the street) while those who didn't (in the hospital, at home, etc.) are ignored, which in this case would include many sick people. As a result, the sample isn't random.

Another example: The survivors of a train wreck attributed their survival to prayer. What wasn't mentioned is that the people who didn't survive also prayed.

Bad Data

Occasionally, even carefully gathered data can have problems. One area that has been extensively studied is longevity, with attention paid to people who live to be older than 100, including "supercentarians," who live to be older than 110. Hopefully, these studies will provide clues for a healthy lifestyle. Traditional predictors of longevity have been no smoking, low salt, low fat diets, high vegetable intake and good social connections.

In July 2019, Saul Justin Newman of the Biological Data Science Institute at the Australian National University, posted a preprint of a paper on the website bioRxiv.org in which he identified a new predictor of longevity: Bad record keeping. (https://www.vox.com/2019/8/8/20758813/secrets-ultra-elderly-supercentenarians-fraud-error). Newman found that in the United States

> "supercentenarian status is predicted by the absence of vital registration. The state-specific introduction of birth certificates is associated with a 69-82% fall in the number of supercentenarian records."

In other words, when states started keeping careful track of birthdates, the number of supercentenarian people dropped. In the U.S., Newman notes,

> "The introduction of complete vital registration in the USA coincided with this rapid increase in lifespan and population size and was expected to result in a large increase in the number of supercentenarian records per capita. From 1880 to 1900, the core survey period for supercentenarians, the US population increased by 150% and average life expectancy around 20%."

However, the introduction of state-wide birth certification coincided with a sharp reduction in the number of supercentenarians. Newman continued,

> "When these states transition to state-wide birth registration, the number of supercentenarians falls by 80% per year, or approximately 69% per capita."

Finally, Newman says,

> "These data reveal that remarkable age attainment is predicted by indicators of error and fraud, including illiteracy, poverty, high crime rates, short average lifespans, and the absence of birth certificates. As a result, these findings raise serious questions about the validity of an extensive body of research based on the remarkable reported ages of populations and individuals."

Longevity researchers may not have paid close attention to an age-old truth: People lie about their age. Wannabe supercentarians are no exception. Teenagers lie about their age to appear older and to get served in bars. Middle-aged people lie about their age to appear younger seem youthful. Senior citizens lie about their age to appear even older than they are so they will look good for their supposed age. Accurate birth records solve this problem, and so it's most noticeable in primitive societies.

Bottom Line

Random sampling has been used to get unbiased information since ancient oracles tossed the I Ching coins to get a random divination. Random sampling allows researchers to accurately estimate certain measurable characteristics of a population. A relatively small random sample will provide accurate information from a large population. For example, a random sample of about 1,500 respondents is 95% reliable and accurate to within 3 percentage points and provides an estimate of the percentage in the population who have a certain opinion, regardless of the population size. Some random samples, like election polls are unreliable, not necessarily because of the polling method, but because of problems inherent in asking someone who they will vote for.

CHAPTER 10

P-VALUES AND PSYCHICS

"As one judge said to another, 'Be just, and if you can't be just, be arbitrary'"
William Burroughs, "Naked Lunch"

Suppose you need to answer a "yes or no" question about a population or a process, based on a random sample. Is a marketing campaign effective or are the sample results due to chance? Is a vaccine effective? Does a fortune teller have psychic powers?

Everett Collection/Shutterstock.com

Madam Zola – Does She or Doesn't She (have psychic powers) ?

Madam Zola, the fortune teller, claims to have psychic powers, also known as ESP (Extra Sensory Perception). A researcher will test her claim. There are two cards, one with a circle and one with a square. The researcher and Madam Zola sit opposite each other with an opaque barrier between them. The researcher holds up a card behind the barrier and Madam Zola attempts to sense the image on the card without seeing it. This is a version of the Zener Card Test, developed by Dr. J.B. Rhine at the Parapsychology Laboratory at Duke University in the 1930s. Madam Zola correctly guesses the square. "Aha!" she says. "This proves that I have psychic powers!"

"Not so fast," says the researcher. "Someone who has no psychic powers and is just guessing has a 50% chance of being correct in one guess, so there's no way to tell with one guess whether or not you're doing better than random guessing."

They repeat the experiment 100 times, and Madam Zola gets 67 correct guesses. Madam Zola claims victory. "On average, a random guesser will get 50% correct, but I got 67% correct," she says. "This proves that I have psychic powers!" How can you tell whether or not Madame Zola was just lucky?

While it's true that on average, a random guesser will get about 50% correct, just like coin tossing, there's variation in random guessing. Some random guessers will do better than 50% (good luck), and some will do worse (bad luck). To decide whether Madam Zola got lucky or whether her result was better than luck, the researcher finds the "p-value" of her result, in this case, the probability that someone who is just guessing will get at least 67 correct out of 100 guesses.

Madam Zola's guesses are presumably a representative sample of her abilities. Was she just lucky? If Madam Zola were guessing at random, her success probability on any guess would be $p = 0.5$. With $p = 0.5$ and $n = 100$ guesses, the probability of getting a result of 67 or better correct guesses, called the p-value, equals 0.0004, or 1 in 2,500. This probability is obtained by using the "binomial distribution", which I'll discuss in Chapter 25. Because her p-value is so low, Madame Zola's result is "statistically significant," better than if she were guessing at random and just lucky, leading the researcher to reject ran-

zendograph/Shutterstock.com

dom guessing and conclude that something else is going on. Note: This does not mean that a random guesser will *never* get a p-value this low, only that it is very unlikely. Such is the case with statistical decisions, where things unlikely to happen just by chance are rejected.

Madam Zola is delighted. She gleefully proclaims, "This is statistical proof that I have psychic powers!" But is it? The hypothesis isn't that Madam Zola has psychic powers but only that she is doing better than random guessing. In addition to psychic powers, there are other plausible explanations for Madam Zola's significant result.

As Madam Zola is leaving the testing facility, the researcher notices a glint on the wall. Upon closer inspection, the researcher discovers a small mirror affixed to the wall in Madam Zola's line of sight. The mirror allowed Madam Zola to get a glimpse of the selected cards, giving her an advantage. She did not have psychic powers. She was cheating, a plausible alternative to random guessing.

P-VALUES

Adapting the legal concept "innocent until proven guilty," which for us means "random until proven otherwise," the burden of proof is on Madam Zola, whom we assume is guessing at random until proven otherwise, statistically speaking. The way to do this is to see how likely it is for the observed result to occur just by chance. The chance probability is called the "p-value". If the p-value is very low, you reject chance guessing and conclude that Madam Zola was doing better than this. There needs to be a p-value cutoff point between "random guessing" and "better than random guessing". This cutoff point is called the "significance level".

SIGNIFICANCE LEVEL

Researchers in all walks of life use 0.05 as a cutoff point to decide whether or not to reject "luck due to random guessing" in favor of a plausible alternative, in this case, that Madam Zola was doing better than chance guessing. This cutoff point is somewhat arbitrary but widely used, so that's what I'll use. Madam Zola's success record had a p-value equal to 0.0004, much lower than the significance level of 0.05, and so the conclusion is that her result was better than luck due to random guessing. This cutoff point between rejecting and not rejecting luck due to random guessing is called the *significance level* of the test. Researchers sometimes don't even mention the cutoff point and just say the result is "statistically significant," or just "significant," whenever the p-value is less than 0.05. In some situations, it makes sense to use a lower cutoff point, but for situations like this, 0.05 is good enough. In order to make a yes-no decision based on observed results, you have to draw the line between yes and no.

Madam Zola's p-value was 0.0004, considerably less than 0.05. As I mentioned earlier, this does not mean that you would *never* get a p-value this low by random guessing, but only that it is very unlikely, even if you're lucky. Statistical logic dictates that if you see something very unlikely under the chance hypothesis, then you reject it in favor of a plausible alternative. In this case, the statistical test picked up the fact that Madam Zola's result didn't occur from random guessing.

Chance Needs Opportunity

Investigating the possible existence of psychic phenomena, called "parapsychology" has been a popular area of research since the 1930's, when a parapsychology lab was established at Duke University. From the Duke University Libraries website:

> "The Parapsychology Laboratory began at Duke in 1930 after Dr. William McDougall invited Drs. J.B. and Louisa Rhine to Durham. At this time, on the coattails of Mesmerism and a public interest in mediumship and other strange phenomena, there was a push in the scientific community to empirically study psychical and paranormal experiences. Thus, parapsychology was born. In the thirty-five years that the Laboratory was at Duke, many words we associate with the paranormal were defined, including extrasensory perception (ESP), telepathy, psychokinesis, pre-cognition, and clairvoyance. J.B. himself was primarily concerned with ESP and many tests of ability were developed, including the famous Zener card test. At its most basic, this test involved a researcher and a participant sitting opposite each other with an opaque barrier between them. The researcher would then hold up a card behind the barrier and the participant would attempt to sense the image on the card sight unseen.
>
> This and other tests quickly caught public attention. About half of the boxes in the Parapsychology Laboratory Records collection are full of correspondence from people around the world who wrote in seeking answers for paranormal phenomena they experienced. Among these are letters from influential people of the time like Timothy Leary, Aldous Huxley, Carl Jung, Harry Houdini, and Upton Sinclair. Some of these, including a particularly mysterious story involving Jung and an exploding knife, can be found in the accompanying digital exhibit."

In addition to the lab at Duke (now in Durham), there are numerous groups around the world, such as the Parapsychological Association and the Berkeley Psychic Institute, who are involved in researching psychic phenomena, hoping that it exists. Similar to the fact that there will be winners of the Mega Millions jackpot when many tickets are purchased, if there are many research projects like this there will be false positives, results that are deemed significant but that are just due to chance. In other words, if you test enough people for psychic powers, some will get lucky and appear to have them even if nobody has these powers.

More Zener Card Guessing

Aida Servi/Shutterstock.com

Suppose that there are 100 Zener card guessing experiments like the experiment in which Madam Zola was caught cheating. Suppose that none of the participants has psychic powers, nobody is cheating, and everyone is guessing at random. Then, using 0.05 as the cutoff point, if the researchers compute each participant's p-value there will be an average of 5 false positives, experiments in which random guessing is falsely rejected in favor of an alternative to guessing. In other words, there will be about 5 random guessers who will be proclaimed to have psychic powers when they are actually just guessing at random. This is a challenge and sets the stage for cherry picking.

Suppose that someone interested in psychic phenomena hears about a significant result from a Zener card guessing experiment and, ignoring the many results that were not significant, offers this as proof of the existence of psychic powers (In the late 20th century such stories were abundant. I know. I was there.) This is a classic example of cherry picking, looking only at results that are significant and ignoring the ones that aren't, without considering false positives due to multiple experiments.

The famous 20th century Swiss psychiatrist, Carl Jung, a brilliant and prolific researcher and writer, and one of the world's greatest cherry pickers, suggested that seemingly significant results in a Zener card experiment demonstrated the existence of "synchronicity," coincidences that Jung thought weren't adequately explained by chance. Jung defined synchronicity as "a hypothetical factor equal in rank to causality as a principle of explanation." In his book, called "Synchronicity: An Acausal Connecting Principle," Jung repeatedly mentions Dr. Rhine's Zener card guessing experiments and cherry picks the results of these experiments. For example, on p. 24 of "Synchronicity," when referring to Zener card guessing experiments, Jung says:

> "Nevertheless, it must be remembered that with Rhine the first series of experiments generally produced the best results, which then quickly fell off. But when it was possible to arouse a new interest in the essentially rather boring experiment, the results improved again. It follows from this that the emotional factor plays an important role."

It has been verified through attempts to replicate these experiments that the "emotional factor" doesn't mean bupkis when you're guessing the picture on a hidden card. It seems that synchronicity was really just another way of saying chance, particularly since there is so much opportunity from multiple testing.

Meaningful and Meaningless

Part of Jung's definition of synchronicity requires that synchronous events be "meaningful coincidences – which are to be distinguished from meaningless chance groupings." Stories about significant Zener card guessing results caught Jung's attention. The *opportunity* for these results, due to random guessing created by multiple experiments, apparently did not. And who is to say what constitutes a "meaningful coincidence" and what constitutes a "meaningless chance grouping?" It's true, however, that extremely unlikely events catch people's attention, especially if they are woven into an interesting story. Most of us have been drawn to such stories, sometimes about lottery jackpot winners. Just remember that both meaningful coincidences and meaningless chance groupings can be caused by chance.

False Positives

One way to debunk false positives is to do further analysis and try to replicate the original results. If a new researcher can't get a significant result using the same methods as the original researcher, then the original result may have been just a random thing. In an August 30, 2021, article in "Live Science," titled "What is ESP?", Isobel Whitcomb discusses this and mentions the lack of replicability as well as non-significant results in situations like Zener card guessing experiments. Whitcomb says,

> "There's a major problem with ESP research: You can't replicate it. 'In science, if you discover something and claim it as a fact, then other scientists following similar procedures should find the same thing,' said James Alcock, a professor of psychology at York University in Toronto. [That] never happens with ESP."

Whether it's ESP or climate change or biotechnology, the popularity of a research subject means that multiple studies will be done. In these situations, the problem of false positives due to multiple tests should be dealt with. In addition to trying to replicate an experiment, you can lower the significance levels (cutoff points) using methods like the *Bonferroni correction* , a conservative method of lowering significance levels to avoid false positives. If you do that, however, you introduce *false negatives*, in which significant results aren't recognized. Replication, if possible, is the best way to guard against false positives.

Fake Psychics Sometimes Use Inside Information to Gain Confidence

Madame Zola isn't the only psychic who was caught cheating. On August 12, 2022, a Washington Post article titled, "Daughter Conned Mother Out of $140 Million With Psychic Scam, Police Say," described how the daughter of an elderly matron "worked with phony psychics to swindle her 82-year-old mother out of art, jewels, and money." (https://www.washingtonpost.com/nation/2022/08/12/brazil-artwork-psychic-scam/)

The story describes how "a rich elderly widow was leaving a Brazilian bank in January 2020, when a psychic approached with a terrible prophecy. Her daughter would soon get sick and die." The mother was suspicious at first, but with her daughter providing inside information to the psychics, she started believing them when they seemed to know personal knowledge of her life that nobody outside the family would know. Over the next two years, the fake psychics, with the aid of the daughter, stole millions of dollars' worth of art, jewelry, and money from the unsuspecting woman, who was deceived by the seemingly psychic abilities of the swindlers, who were given confidential information by the daughter.

Finally, police who were investigating elder abuse uncovered the scam and arrested the daughter and four alleged psychics. The story states that "during a raid of the home of one of the purported psychics, police found 11 paintings under a bed, Reuters reported. At the bottom of the pile, they discovered 'Sol Poente,' which investigators said was valued at about $48.5 million." The story concluded that more arrests are expected.

WAVING THE TOWEL

You're walking down a residential street in Berkeley, California, when you see a man standing in front of a house and waving a towel. You ask him what he's doing. He says that he's keeping the elephants away. You point out that there aren't any elephants in the area. He says, "See, it's working."

Consider each day a "success" (no elephants) or a "failure" (elephant shows up). You observe the area for 50 days and there are no elephants - all successes. Do you conclude that waving the towel is effective at keeping the elephants away? There are a few obvious reasons for the towel-waver's success. First, there are no elephants in Berkely, California; there's not even a zoo. In order to find a p-value, you have to first find the probability that a random elephant would show up if the person were not waving a towel, but since there aren't any elephants in the vicinity, this isn't possible. There's no way to test the claim without some elephants having the potential to approach the clown. The towel waver's conclusion could be attributed to what social scientists call "confirmation bias," in which the person is claiming success because there's no evidence to the contrary, a weak argument. It seems that the way out of this dilemma is to bring in the elephants.

Bottom Line

P-values can distinguish between events that happen by chance vs. some other plausible explanation. The "significance level" of such a test draws the line between "significant" (not caused by chance) and not significant (caused by chance), with the standard (arbitrary) significance level equal to 0.05. An event is considered "significant" if the probability that it happens due to chance is less than 0.05. Multiple tests or comparisons must be treated with care because of the opportunity for false positives. In such situations, replicating the tests, if possible, is the best way to weed out false positives. Otherwise, statistical methods to lower the significance level must be considered.

CHAPTER 11

PAUL THE OCTOPUS PICKS WINNERS OF SOCCER MATCHES

"A sports expert is the guy who has the best alibis for being wrong."
- Jimmy Cannon, legendary sports writer

A gambler was hopelessly hooked on football betting. Nothing else interested him. Unfortunately, he lost almost every bet he made. Finally, even his bookie felt sorry for him. "You lose all your bets," said the bookie. "Why don't you bet on hockey instead of football?"

"Hockey?" said the gambler in dismay. "I don't know anything about hockey!"

Paul the Octopus

Paul the Octopus was born in 2008 and lived in the Sea Life Center in Oberhausen, Germany. Paul became famous for successfully predicting World Cup soccer matches in the years 2008-2010. According to a Wikipedia article (https://en.wikipedia.org/wiki/Paul_the_Octopus),

"Paul's keepers would present him with two food-containing boxes decorated with the flags of the teams in an upcoming match. Whichever box Paul ate from first was considered his prediction for which team would win the match."

ennar0/Shutterstock.com

It was Paul's success record that made him famous. According to the same article,

> "His keepers at the Sea Life Centre in Oberhausen, Germany, mainly tasked him with predicting the outcomes of international matches in which the German team was playing. Paul correctly chose the winning team in four of Germany›s six Euro 2008 matches, and all seven of their matches in the 2010 World Cup—including Germany›s third-place play-off win over Uruguay on 10 July. He also correctly chose Spain as the winner of the 2010 FIFA World Cup final. In all, Paul amassed an overall record of 12 correct predictions out of 14: a success rate of approximately 85.7%."

There are a few details that aren't mentioned in the article. For example, in World Cup matches, some teams are better than others. Maybe Paul's keepers put better food in the box representing the favorite. Maybe Paul's keepers didn't pay close attention if Paul happened to get food from both boxes and then said Paul picked the winning team after the match was over. Paul died of natural causes in 2010, so we don't have him around to try and replicate his results.

Let's assume that Paul was guessing at random and had a 50% chance of being correct on each guess. Then, the number of correct guesses in Paul's total of 14 predictions has a binomial distribution with n = 14 and p = 0.5 (see Chapter

25). If X = number of correct guesses, we can find the p-value, $P(X \geq 12) = 0.0065$, or about 1/155. This is a significant result. In other words, our conclusion is that, like Madam Zola, Paul the octopus was doing better than random guessing. Unlike Madam Zola, Paul couldn't have been cheating with a mirror or any other device, since he was an octopus, and since the outcomes of the matches weren't known until after Paul made his picks. The p-value, expressed as 1/155, means that it will take an average of 155 attempts at picking 14 soccer matches like this, by Paul or some other prognosticator who is guessing at random, before a result as good as Paul's occurs. One explanation, as always, is opportunity. If there are lots of similar situations, then Paul's results are not so surprising. Remember, it takes lots of losers to produce a winner, even when we are looking at a prognosticating octopus. Sure enough, the article further stated:

> "Some other German oracles did not fare so well in the World Cup. The animals at the Chemnitz Zoo were wrong on all of Germany's group-stage games, with Leon the porcupine picking Australia, Petty the pygmy hippopotamus spurning Germany›s apple-topped pile of hay (instead of Serbia), and Anton the tamarin eating a raisin representing Ghana. Mani the parakeet of Singapore became famous for correctly predicting the results of all four quarter-final matches. Mani contradicted Paul by picking the Netherlands to win the final, resulting in some media outlets describing the game as an "octopus-versus-parakeet showdown."

Just as a lottery jackpot winner exists because of opportunity created by the losers, the success of Paul the octopus was partially due to the other animal prognosticators who may not have done so well but were part of the group that produced a winner. Paul was the one who got lucky. Ten years later, Paul's successor, Rabiot, the giant octopus, came on the scene in Japan. According to an ABC news report by Molly Hunter, titled, "Octopus killed for food after correctly predicting three World Cup games," "Rabiot, the World Cup-predicting octopus, has died. More precisely, he's been turned into sashimi, reports Japanese newspaper Mainichi."

The story goes on to say that,

> "Rabiot was a Giant Pacific Octopus caught off the coast of Obira, Hokkaido. According to Mainichi, fisherman Kimio Abe, 51, named the mollusk on June 19. Abe then placed his new invertebrate in a paddling pool containing three boxes with food, one

> representing Japan, another representing its opponent and a third representing a draw, according to reports. In the experiment, the octopus tipped its tentacles towards the right boxes, thus predicting a Japanese win against Colombia, a draw with Senegal and a loss to Poland. All three predictions were true. 'I'm glad that all the forecasts turned out correct and Japan moved on to the knock-out stage,' Abe told Mainichi in an interview published before the team's loss to Poland."

Unfortunately, the day after Rabiot's predictions, Mr. Abe, in a classic case of kill the prognosticator, ended Rabiot's brief career and turned him into sushi. At the same time, ABC news also reported that Achilles, a deaf, white cat, who also predicts winners of soccer games by picking a food bowl representing a team in a matchup, became the official prognosticator for the Russian World Cup.

A CLEVER ECONOMIST PREDICTS STOCK MARKET CRASHES

Levent Konuk/Shutterstock.com

As further evidence of people only noticing "meaningful events," as Jung puts it, there's a famous TV personality and economist who has correctly predicted every stock market crash for the past thirty years. Every week he predicts a stock market crash. He's wrong most of the time, but every so often, there is a stock market crash, and he correctly predicts it (because he always predicts crashes). Nobody pays attention to his incorrect predictions, but people remember meaningful events like stock market crashes. They say, "That's the guy who predicted the stock market crash of 2008!" You don't hear people say, "That's the guy who didn't predict that the stock market wouldn't crash in 2012!"

The economist's predictions are reminiscent of Paul the Octopus.

Sports Services – You only pay when you win

The increased popularity of sports betting has created a cottage industry of sports services and touts. Sports services are companies that sell picks and information, supposedly to help you be a sports betting winner. Sports services often have different categories of picks. For example, a service may have "key picks," "early picks", "late picks," "five star picks," "gold label picks," "technical picks," etc. Each category of picks favors different sets of games, so that opportunity is created for at least one category to have a winning record. If the "key picks" record is 65% winners, the sports service will advertise, "Key picks – 65% winners!" without mentioning the many categories with losing records.

Another scam is to say, "We only charge you for winning picks! If our pick loses, you pay nothing." Sounds good, right? If the service's pick loses, it's true that you don't pay them, but if you bet on it, which is why you're doing this in the first place, you pay the casino when you lose. Suppose the sports service doesn't have any expertise and just tosses a coin to make their picks. Then, just by chance, they'll win an average of 50% of their picks. Suppose you pay the sports service $100 for each winning pick and nothing for losing picks. If the sports service wins 50% of their picks, they'll make an average of $50 profit per pick for every customer. The icing on the cake is this: You'll be paying $100 a pick to bet on teams someone chose by tossing a coin!

You can always win 50% of the time

Suppose that the sports service I just discussed always tosses a coin to pick a team, A or B, to win a game. Suppose, for example, that the probability that team A beats team B equals 0.75, and so the probability that team B beats team A equals 0.25, and that the sports service, who doesn't care about these probabilities, tosses a coin to decide their pick. Then the chance that the sports service wins its pick equals: $0.5*0.75 + 0.5*0.25 = 0.5*(0.75 + 0.25) = 0.5$.

The sports service will have the same result regardless of the probability that team A beats team B. So, in repeated coin toss picks, even though the sports service will have winning streaks and losing streaks, they will average 50%

winning picks, regardless of the strengths and weaknesses of the competing teams. In fact, since picks made by coin tossing will have winning and losing streaks just due to chance, the sports service will advertise its winning streaks and get new customers who are impressed with the streaks without realizing that it comes from coin tossing.

A Winning Strategy?

A sports gambler claims that if he hops around the room and quacks like a duck before each game, then he has greater than a 60% chance of winning his bet. The gambler's friend, the statistician, tests the claim by observing the results of 50 such bets in which, before each bet the gambler hops around the room and quacks like a duck. The gambler wins 26 bets out of 50, a 52% win percentage. Comparing this result to 60%, the statistician points out that 52% is less than the gambler's claim of 60%, and so he rejects the gambler's claim and says he is not even lucky. The gambler says that his method doesn't work if someone is watching him. The statistician says that the gambler is like a quantum particle in which the wave function collapses as soon as it is observed.

Later, the statistician observes the results of 50 bets where the gambler *doesn't* hop around the room and quack like a duck. The gambler wins 25 out of 50, or 50%. The gambler has the last laugh, saying, "See, I did better when I quacked like a duck." The statistician points out that the difference isn't significant, but the gambler isn't listening. He improves his technique and starts shrieking like a baboon before making his bets.

The Parrot

A gambler bought a parrot who had retired from a successful career of picking winners of baseball games and was alarmed to hear the parrot curse at every opportunity. The parrot's new owner tried everything, from covering the parrot's cage with a blanket to withholding food but there was still non-stop cursing. Finally the frustrated owner took the parrot out of its cage and put it in the freezer. After an hour the owner removed the shivering parrot from the freezer and put it back in its cage. The parrot looked shaken. It didn't say anything for two days. Finally, as the parrot's owner was walking past its cage, the parrot looked at him and said, "Just one question. What did the chicken do?"

Bottom Line

Some animals who pick the results of sporting events appear to be skilled because of their high win percentage. Upon further checking, it seems that their supposed skill is the result of false positives caused by a large number of animal prognosticators. In other words, a few of them got lucky.

If you randomly guess the winner of a game with two teams or contestants, you will be correct about 50% of the time. Unscrupulous sports services may use this fact to make a profit even when a gambler who uses the service only pays for winning picks.

CHAPTER 12

HOW MUCH CAN YOU LOSE WITHOUT BEING CHEATED?

A businessman raised $40 million in investments for his company. The company was supposedly on the verge of obtaining a patent for a lucrative new dental product. Such a product existed, but the businessman had no connection to it. He was a swindler. The businessman used $10 million to buy homes, condos, and luxury cars. He gambled away the remaining $30 million betting on sporting events at an offshore casino. Eventually, he was apprehended. He declared bankruptcy and was sent to prison.

The swindled victims retained a law firm that specializes in recovering funds from bankruptcy cases. The law firm got a court to seize the swindler's homes and cars. They filed a lawsuit against the offshore casino on behalf of the swindled investors. The lawsuit alleged that the offshore casino must return the money that the swindler lost gambling because the casino was either a) laundering money for the swindler, or b) defrauding the swindler by not giving him the proper odds on his bets.

The lawsuit claimed that if the casino had given the swindler proper betting odds he couldn't possibly have lost $30 million over a 1.5 year period. The attorneys representing the offshore casino retained me to show that in fact it was possible for a gambler to lose $30 million in 1.5 years making the types of bets he was making. The swindler was making 4-team parlay bets on various sports. He made an average of ten bets per day at $20k each, amounting to $200k per day and resulting in $110 million in betting action, or "churn" over a 1.5 year period. How much, on the average, will a gambler lose when he bets $110 million in relatively small increments over time?

Answer: It depends on the bets. In particular, on the "house edge," which, as we discussed in earlier chapters, is the negative of the gambler's expected value and represents the average percent profit made by the casino.

Let's review roulette as a simple example. A roulette wheel has 38 sections labeled 1 through 36, along with 0 and 00. Suppose you bet on #17. Then your chance of winning equals 1/38 and your chance of losing equals 37/38. Casino payoff odds on this bet are 35 to 1. In repeated $1 bets, by the law of averages a gambler will win about once in 38 bets ($35 payoff) and lose about 37 in 38 bets ($37 loss) for a net loss = (35-37)/38 = 0.053 = 5.3% = House Edge for single number roulette bets. Equivalently, the gambler's EV equals -0.053. If the gambler had been making roulette bets, he would have lost an average of 0.053*$110 million, or about $6 million for $110 million bet, assuming that he was betting in relatively small increments and didn't go broke after a losing streak. If the gambler was making roulette bets, it's very unlikely that he would have lost $30 million making the bet sizes that I've described.

A "parlay" bet is popular among sports bettors who like high payoff odds. Parlay bets are bets on the outcomes of more than one game or event in which all parts of the bet must win in order for the parlay bet to win. Here's an example of a four-team football parlay, the type of bet the swindler was making.

> 49ers beat Seahawks, Steelers beat Browns, Chiefs beat Raiders, Rams beat Cardinals.

The gambler wins the 4-way parlay bet only if all 4 picks win. Assuming that the different parts of the parlay are independent, the chance of winning is the product of the chances of each individual part winning. (Independence is not always the case but that doesn't affect the current discussion.) For American football and basketball parlay bets, point spreads make the matchups in each leg of a parlay bet 50-50 propositions (unless you're a skilled player), so the theoretical probability of a 4-way parlay with football and/or basketball picks is ½ to the 4th power, which equals 1/16. For games with odds instead of point spreads, like baseball and soccer, the computations are similar. You convert the odds into probabilities and then multiply.

The standard payoff odds on a 4-way parlay with points spreads are 10 to 1. Using P(win) = 1/16, the gambler's EV (expected value) when making a 4-way parlay bet is then

EV = (1/16)*10 – 15/16 = -0.3125, giving the casino a 31.25% edge.

For a total of $110 million in bets, the gambler would expect to lose about 0.3125*$110 million = $34.375 million, more if he went broke first. This is consistent with the losses of the Canadian swindler.

That was not the end of it. The attorneys for the swindler's victims didn't accept the argument about house edges for parlay bets, partially because of the poor record keeping of the offshore sportsbook. The attorneys subsequently complained to the attorney general of the country where the offshore casino was located (The "offshore casino" was an office where employees sat around with laptops taking online sports bets). The attorney general impounded $8 million of the casino's money until the lawsuit was settled, the remainder of the $30 million being unavailable. In subsequent discussions, attorneys for the victims and the owner of the offshore casino agreed to a $6 million settlement. The attorney general's office kept the remaining $2 million as an "administrative fee." And that was the end of that.

Bottom Line

It's surprising how much an unskilled gambler can lose.

CHAPTER 13

LUCK VS. SKILL

"You've Got to Ask Yourself One Question:
'Do I Feel Lucky?' Well, Do Ya, Punk?"
Clint Eastwood as "Dirty Harry"

What is Luck?

According to the Oxford Languages dictionary: "Luck is success or failure brought by chance rather than through one's own actions." Luck is based entirely on chance. What are "success" and "failure?" One person's success is another person's failure. Sometimes we can agree, but not always. In a casino game of chance, like craps or roulette where we can compute a gambler's expected value, we could say that good luck is getting a higher average winnings than the EV and bad luck is getting a lower average winnings than the EV. The most important part of the definition is that luck, however you define it, is due to chance. If you have success due to skill, or due to cheating, that's not luck.

We feel lucky or unlucky based on things that happen every day. It doesn't have to be as spectacular as winning the lottery or getting run over by a bus, but, in a sense, it has to be meaningful. You might think you're lucky if you wake up and it's a beautiful day or if you got a raise. You might feel unlucky if you catch a cold or if you lose money at a casino.

Gorodenkoff/Shutterstock.com

There are superstitions. Different cultures have lucky numbers (7 in the U.S. and India, 3,5, and 8 in China) and unlucky numbers (13 and 666 in the U.S., 4 in China, 17 in India, 43 in Japan, and 39 in Afghanistan). There are superstitious events like stepping on a crack in the sidewalk (unlucky), walking under a ladder (unlucky), meeting up with a long-lost friend (hopefully lucky), and astrological readings that can go either way. Gamblers try and find lucky slot machines and avoid unlucky ones.

Luck is All There Is

Craps is a game of chance, no skill, so if you end up a winner, it's because you are lucky. We analyzed bets on "7", that is, that 7 would come up on the next roll of the dice, which has probability equal 1/6 and saw, by the law of averages, that if you keep making this bet, with payoff odds 4 to 1, you will end up losing at the rate of 17 cents per dollar bet. If you're playing a game of chance like craps, luck is all there is. And, by the law of averages, luck eventually disappears. For the same reason that there are jackpot winners in the Mega Millions lottery, if there are enough craps players, *someone* will be a lucky winner due to the opportunity presented by many players. But even winners will start losing if they keep playing.

Some gamblers have a more complicated definition of good luck than just winning money. For example, the quote from a Las Vegas gambler, "I hope I break even this week, I need the money," indicates that a gambler who's been through tough times might consider himself lucky if he just breaks even. A veteran craps player might feel lucky if she loses, but not as much as the 17 cents per dollar that she's *supposed* to lose. Another gambler might attribute losses to entertainment costs and feel good about the whole thing.

Luck in Everyday Life

I first learned about luck when I was six years old. I had just learned to ride a bike. I was riding down a hill when I suddenly found myself going faster than I could manage. I ended up in the bushes, with numerous scrapes and bruises. When I got home, my mother said, "You're lucky you didn't get killed." My mother had a way of interpreting every non-fatal accident as good luck. Was she right? I didn't feel lucky. Maybe there's data showing that the average child's bike crash is worse than mine. In a sense, my mother was right. If I did better than the average kid who crashes their bike, I guess I was lucky, or maybe skilled, which is different from luck. Or maybe only a few young bike riders would crash in this situation, and if they did, their injuries would be less severe than mine. So maybe I was unlucky. Or maybe my mother was thinking about the conditional probability of getting killed given that I got in a nasty accident to begin with.

Suppose you find accident data for young bike riders. Divide the total number of accidents by the total number of bike rides to get "accidents per ride." This crudely estimates your chance of having an accident if you ride your bike. For example, suppose the average young bike rider has about 1 accident every 25 rides. Suppose you've recently gone on 30 rides. Then, based on this data, it wouldn't be surprising for you to have an accident. Of course, some kids are better bike riders than others and some take more risks than others, so, in addition to luck, skill is a factor. I could have prevented a crash by not riding down the stupid hill in the first place. If I had stayed in bed all day, I would never have had an accident. Or maybe I should have practiced more before riding down the hill.

Sporting events and chess (maybe chess is a sporting event) have both luck and skill. If two professional golfers, two chess players, or two football teams of equal ability, compete against each other (or any two contestants equalized by odds or point spreads), luck becomes a factor. The results are difficult to predict. That's why the contests are exciting, and that's why people bet on them. Maybe the golf ball took a bad bounce or maybe a pass was deflected and became an interception. Or maybe the referee made a bad call. Even games of skill are subject to uncertainties caused by chance. When you say that you're lucky to have such a good job or to be with such a great partner, there's probably more involved than luck. Or maybe you make every decision by tossing a coin. When you say that you're lucky to have such nice parents, is that caused by chance? You can't have skill in picking your parents. You don't have a say in picking them at all. It's not your fault. It's the result of genetics and survival of the fittest and whatever other laws of nature are involved, including chance.

Roulette

Like craps, roulette is a game of chance in which no skill is possible. If you play roulette repeatedly, betting, say, on red, in which the probability of winning equals 18/38, and with 1 to 1 payoff odds, you will eventually lose, with your losses averaging 5.3 cents per dollar bet. If you end up a winner, it's due to luck. Or is it? We discussed Dr. Richard Jarecki, who made a fortune betting on roulette wheels for which he had collected data and discovered imperfections. These roulette wheels were biased and didn't spin randomly. Dr. Jarecki used statistics and took advantage of faulty equipment. He might have considered himself lucky to find the faulty equipment, but, in a sense, his success was due to skill, not luck.

Who's a better Othello player, my friend or me?

My friend and I used to play the board game, Othello, every week. My friend claimed that he was better than me, and I claimed that I was better than him. How does one decide? First, collect some data. In 100 games, spread out over a few months, I won 67 and my friend won 33. I told my friend that my win record of 67% proved that I'm better. He said that I was just lucky. The way to decide between luck and skill is to compute the p-value for this result. Assume that I'm lucky until proven otherwise. If that's the case, then I have a 50% chance of winning a game and so does my friend, and my total wins in 100 games has a binomial distribution (discussed in Chapter 25). If X = number of wins in 100 games, then the p-value equals:

$$P(X \geq 67) = 0.0004, \text{ or about 1 in 2,500.}$$

This is a significant p-value. In other words, it would be very unlikely to have a record like this just by guessing. So, statistically speaking, my success record was not just due to luck. I was more skilled than my friend.

Sports Betting - What's Luck Got to do With It?

Games like craps, roulette, and lotteries are easy to categorize as games of chance because the player is essentially betting against random numbers: A player in such a game has no strategy that does better on average than random guessing. Blackjack involves both chance and skill. Even though dealing cards from a deck is random, there are strategies based on conditional probabilities arising from cards being dealt that can give a player an advantage. Some app games may involve skill, but how can you tell the difference between luck and skill? If there is no skill, a player's results will be due to chance, with the winners being the ones who got lucky. This has more than academic importance. Wagering on "fantasy" sports and other events related to sports events is legal online but in some states these wagering activities must be skill-based. This means that game developers need an expert opinion regarding luck vs. skill.

I analyzed such a game in which contestants wager on the results of sporting events. A contestant either wins or loses each wager. Whether you're wagering on sports or other events, like the stock market, it's sometimes not clear whether skill is possible. Although you can show mathematically that there's no skill in craps and roulette, you can't use the same argument to decide whether picking the result of a sporting event is a game of chance or a game of skill. For example, you can research the history of teams and maybe get some insights. You can look at player data and analyze the outcomes of repeated play. Maybe you have knowledge that give you a better chance of winning than just being lucky. To decide between luck and skill, we use p-values.

Assume that a contestant is lucky until proven otherwise and has a 50-50 chance of winning a wager. Then find a p-value for the results. In other words, assume that the result of each wager is random, like the result of a coin toss, unless proven otherwise. If, under this assumption, some contestants do significantly better than luck, you conclude that this is a skill-based game. (Assume there is no cheating.) With this in mind, call a *skilled* contestant someone who has a significantly higher win percentage than chance (luck) allows. An *unskilled* contestant is someone whose win percentage is within the range of luck.

P - VALUES

You reject the "luck" hypothesis and conclude that a contestant is skilled if there is an extremely low probability that the contestant could have achieved their observed win percentage for the number of wagers they made just by guessing. If this is the case, conclude that something other than chance, namely skill, is present. I found contestants who had low p-values and whose performance was better than just being lucky and concluded that this was a skill-based game. This doesn't mean that chance and luck don't exist, but rather that a contestant can develop skilled strategies.

MULTIPLE COMPARISONS

For an analysis like this, you must conduct many tests, one for each contestant, to determine if that player is skilled. Each player has his or her p-value, determined by the win-loss record for the number of predictions made. If sufficiently low, the p-value indicates statistically that the player is skilled, not just lucky. Since there are multiple statistical tests, sometimes called multiple comparisons, you should do something like adjust the p-value cutoff point that distinguishes skill from luck to account for the opportunity for false positives. As with the tests for psychic powers, if you test enough sports wagerers, some will appear skilled just due to chance and the opportunity presented by so many bettors. To avoid this possibility, I made sure that the p-value cutoff between luck and skill accounted for the total number of players. (There are methods for doing this, but I won't go into it here.)

In the wagering data, 32 out of 1,980 contestants had win percentages for the number of predictions they made that were so high as to rule out luck. The other contestants were either within or below the bounds of chance. These results are consistent with a skill-based game.

The contestant with the lowest p-value in Table 1 below had 522 wins in 793 bets, for a 65.8% win percentage. A 65.8% win percentage in 793 predictions just due to luck has probability equal to 0 rounded off to 30 decimal places The 2nd lowest p-value equals 0.00000000000370348197, or about 1 in 270 billion. The 5 lowest p-values are in Table 1.

TABLE 1.

player	wagers	wins	losses	win %	p-value
1	793	522	271	65.8%	0.00000000000000000000
2	574	369	205	64.3%	0.00000000000370348197
3	373	241	132	64.6%	0.00000000895616669716
4	403	257	146	63.8%	0.00000001760864176692
5	168	119	49	70.8%	0.00000003323155139956

The statistical analysis of the game showed that even when conservatively correcting for false positives, 32 contestants did better than they would just by being lucky. Someone can get lucky and have a higher than 50% win percentage for a while, but in repeated play, it's more difficult. Skill will lead to a long-run win percentage that cannot be attributed to luck. The analysis showed skilled contestants. Luck has its limits. You can show how to detect when someone goes outside these limits.

Bottom Line

Using p-values, you can test whether a gambler is skilled (significant win-loss results in repeated play) or just lucky (results are not significant). However, if you're testing numerous contestants, opportunity rears its smiley face, and you must account for false positives. If there are some contestants who are skilled at picking winners, as measured by p-values, you can conclude that the contest itself is a game of skill, even if some contestants don't develop the skills that are available to all.

CHAPTER 14

LOADED DICE PROVIDE POSITIVE EXPECTED VALUE

What if betting on 7 in craps had positive expected value? There are two ways to accomplish this: 1) Increase the probability of winning (use "loaded" dice), or 2) Increase the payoff odds. The casino is not about to do either of these things, so it's up to you. (Warning: don't try this at home).

Suppose you are an unscrupulous gambler who has loaded or weighted dice for which P(7) = ¼ instead of 1/6. After putting "loads" in the dice, you estimated probabilities by rolling the loaded dice many times and finding the fraction of times that 7 came up. The law of averages works for loaded dice as well as ordinary dice, but you don't have the nice combinatorial argument that works for ordinary, evenly-balanced dice (As we discussed in the Craps chapter, there are 6*6 = 36 equally likely combinations if you roll a pair of evenly balanced dice, with 6 ways to get "7", so, since the dice are evenly balanced, P(7) = 6/36 = 1/6. If the dice are not evenly-balanced, the computation doesn't work.)

Suppose that you sneak your loaded dice into a casino and bet on 7, with the usual casino payoff odds of 4 to 1. Then your expected value equals (1/4)*4 – (3/4)*1 = 1/4 = $0.25. You have a bet with positive expected value! If you bet prudently, you will have a long run profit, at least until casino security discovers that you have loaded dice and throw you under a bus, literally.

Consider the following scenario. Say you have $1,000 to bet. Since your bet has positive expected value, the law of averages says you will win in the long run. So you might as well bet everything you have, namely $1,000, right? Wrong. Let's say you win the first bet. At payoff odds of 4 to 1, you now have $4,000 plus your original stake, for a total of $5,000. The expected value doesn't change. It's still 25 cents per dollar bet. Might as well bet it all again. Unfortunately, on the second bet, "9" comes up and you lose. Since you bet everything, you're tapped out, bankrupt. In fact, for any strategy in which you make large bets like this, especially making large bets to cover your losses like the double-up strategy, even if you have positive expected value, you will go broke. It's not true that positive expected value guarantees that you will make a profit. You must use a prudent betting system, like the Kelly system.

THE KELLY SYSTEM FOR LOADED DICE

For simple situations where you know or can estimate the probabilities of winning, p, and losing, q, and with payoff odds, b to 1, the Kelly system says to bet the fraction of your bankroll equal to p – q/b. With loaded dice, if you bet on "7" in a craps game with payoff odds 4 to 1, b = 4, p = ¼, q = ¾, and so the Kelly system says to bet the fraction ¼ - (3/4)/4 = 1/16 = 6.25% of your bankroll. So instead of betting it all, if you start with $1,000, bet $1,000*6.25% = $62.50. Then, if you win the first bet, you would win 4*$62.50 = $250, bringing your bankroll to $1,250. On your second bet, you would bet $1,250*6.25% = $78.13. Instead of going broke if you lose this bet, you would have $1,250 - $78.13 = $1,171.87, a profit, so far, of $171.87 over your initial bankroll of $1,000. For your third bet you would wager $1,171.87*6.25% = $73.24. And so it goes. Even though your winnings increase slowly, by betting less after you lose, you are, in a sense, protected from going broke.

Suppose your loaded dice are confiscated and you are stuck using ordinary, "fair" dice to make bets. If you bet on "7" without cheating, the payoff odds are 4 to 1 and the chance of winning equals 1/6, so the Kelly strategy says to bet the fraction 1/6 – (5/6)/4 = -1/24 of your bankroll, which is impossible, since you can't bet a negative fraction of your bankroll. In other words, if you have a situation with negative expected value, the Kelly system says don't bet.

Bottom Line

Loaded dice can be used as a metaphor to help understand bets with positive expected value where you can't use a combinatorial argument to compute probabilities and where you must use empirical results instead.

CHAPTER 15

THE CRYPTO KING CONS HIMSELF

"When degenerates lose a bet, they don't double-up to get even.
Who wants to be even? Degens triple-up."
– Michael "Roxy" Roxborough, former head Las Vegas oddsmaker.

Disgraced crypto king Sam Bankman-Fried presumably learned about probability and expected value when he was an undergraduate at MIT and perhaps earlier, maybe in high school. According to some reports, he often made expected value computations when pondering risky decisions, going so far as to take great risks when he perceived that there was positive expected value, "plus EV", as he called it. In his best-selling book about cryptocurrencies, "Number Go Up," investigative reporter Zeke Faux describes conversations he had with Bankman-Fried. In one conversation, Faux says,

> "Returning to the framework of Expected Value, I asked him [Bankman-Fried] if the decisions he made were correct. 'I think I've made a lot of plus EV decisions and a few very large bonehead-ed decisions,' he said. 'Certainly, in retrospect, those very large decisions were very bad, and may end up overwhelming everything else.'"

It's important to note that Bankman-Fried talks about plus EV as if it were a mathematical truth, whereas many of his calculations seem to have been based on estimated numbers and were possibly inaccurate. Faux continues,

> "The way Bankman-Fried later described this decision reveals his attitude towards risk. He estimated there was an 80 percent chance the exchange would fail to attract enough customers. But he said that one should always take a bet, even a long shot one, if the expected value is positive, calling this stance 'risk neutral.' But it actually meant he would take risks that to a normal person would sound insane. 'As an individual, to make a bet where it's like, I'm going to gamble my $10 billion and either get $20 billion or $0, with equal probability, would be madness,' Rob Wiblin, host of an effective altruism podcast, said to Bankman-Fried in April. 'But from an altruistic point of view, it's not so crazy.' 'Completely agree,' Bankman-Fried replied.'"

The thing is, it's crazy from *any* point of view, even from an "altruistic" one, whatever that means, unless it's okay for effective altruists to take bigger risks than ordinary people. Maybe, since they're going to give their money away at some unspecified time after they're rich (after they die?) it doesn't really matter if altruists go broke? And if you're going to base your altruism on risky bets, maybe you should use random altruism instead, like the King did with his lottery in Chapter 1.

Look at the bet: "You have $10 billion and can either get $20 billion or $0, with equal probability." Let's assume that Bankman-Fried means that after the coin is tossed (equal probabilities), you will have either a $20 billion profit or you will lose your $10 billion. This interpretation has expected value equal to $20 billion*0.5 - $10 billion*0.5 = $5 billion, a bet that has positive expected value (plus EV).

As we have seen, finding a bet with positive expected value is the first step, however, you must also have a betting strategy. What appears to have been SBF's betting strategy is this: When you find a bet with positive expected value, bet everything you've got, or more than you should, in any event. If you either make a $20 billion profit or go broke, the payoff odds are 2 to 1. Since the probability that you win or lose equals 0.5, with the Kelly system you should bet the fraction .5 - .5/2 = 0.25 of your bankroll, so instead of betting $10 billion, you should only bet $2.5 billion, still a lot, but now you're protected from quickly going broke. Maybe SBF thinks the Kelly system is only for squares.

There's another problem, which I mentioned earlier: Even though SBF behaved as if he was living in an enormous casino with clearly stated bets, success probabilities and payoff odds aren't clearly stated in the outside world. Most of his expected value computations seem to have been based on estimates, which can be inaccurate, even in the world of crypto gambling. Even though SBF stayed solvent for a while (until he tapped out), he seems to have been cavalier about assuming that his estimated odds and probabilities were accurate. His biggest mathematical mistake seemed to be not following a sensible betting system like the Kelly system, which would have lowered some of his risks and maybe helped avoid the slippery slope that led to his demise. The slippery slope seems to have been made more slippery by making bigger bets to avoid disaster, namely the double-up strategy, leading to borrowing from customers' investments to pay off debts, a situation where math problems yield to legal ones.

Bottom Line

It's easy to broke, even with positive expected value, if you don't use a sensible betting system. When the stakes are high, going broke is not a pleasant outcome. Then again, neither is prison.

CHAPTER 16

BAYES' RULE – A DIAGNOSTIC TEST AND RESEMBLANCE STEREOTYPING

Cards

If you play craps or roulette, you have sequences of independent events. The law of averages waits in the wings, like a predatory bird. But it doesn't have to be that way. Life can be interesting. Walk over to the blackjack table, where things change as cards are dealt from the deck. The probability that a particular card is dealt depends on which cards were previously dealt. Suppose a card

pics five/Shutterstock.com

is dealt from a well-shuffled deck. Since there are 4 aces and 13 spades in a deck of 52 cards, using the rules of probability, we get:

P(Ace) = 4/52, P(Spade) = 13/52

Suppose that two cards are dealt from the deck. What's the probability that the second card is an ace "given" that the first card is an ace? This is called a conditional probability because the probability that the second card is an ace depends on the result of the first card dealt. In a sense, you're updating probabilities based on new information (the result of the first card). In formulas, conditional probability is denoted by a vertical line:

P(2nd card Ace | first card Ace) = 3/51, because if the first card is an Ace there are only 3 aces and only 51 cards left in the deck

P(2nd card Ace | first card is *not* Ace) = 4/51 (because if the first card is not an Ace there are 4 aces and only 51 cards left in the deck)

Also,

P(2nd card spade | first card spade) = 12/51

P(2nd card is not spade | first card spade) = 39/51

P(2nd card spade | first card is *not* spade) = 13/51

The rules of probability apply to conditional probability the same as for regular probability except restricted to the outcomes on which you're conditioning. For example, if events are equally likely, then the probability of an event is the fraction of ways the event can happen in the set of *conditional* outcomes.

On the other hand:

Toss a coin twice.

P(H on 2nd toss | H on 1st toss) = 0.5

P(H on 2nd toss | T on 1st toss) = 0.5

For coin tossing, conditional probabilities are the same as unconditional probabilities. This is one way of defining *independent* events. The coin has no memory of what happened on a previous toss. Same as for rolling dice, spinning roulette wheels, and lottery drawings. But not cards.

Multiplication Rule for Conditional Probability

Here is the multiplication rule for conditional probabilities:

For any events A and B,

P(A and B) = P(A|B) * P(B) = P(B|A) * P(A)

If A and B are independent, then P(A|B) = P(A), and P(B|A) = P(B), and so

P(A and B) = P(A) * P(B), which is the multiplication rule for independent events.

Two cards are dealt from a well shuffled deck. Then P(both cards are Aces) = P(first card is Ace) * P(second card is Ace | first is Ace) = 4/52 * 3/51 = 0.0045

Bayesian Analysis – Diagnostic Test for a Rare Disease

Updating data by using conditional probability is sometimes known as "Bayesian" analysis, named after Thomas Bayes, an English statistician and minister

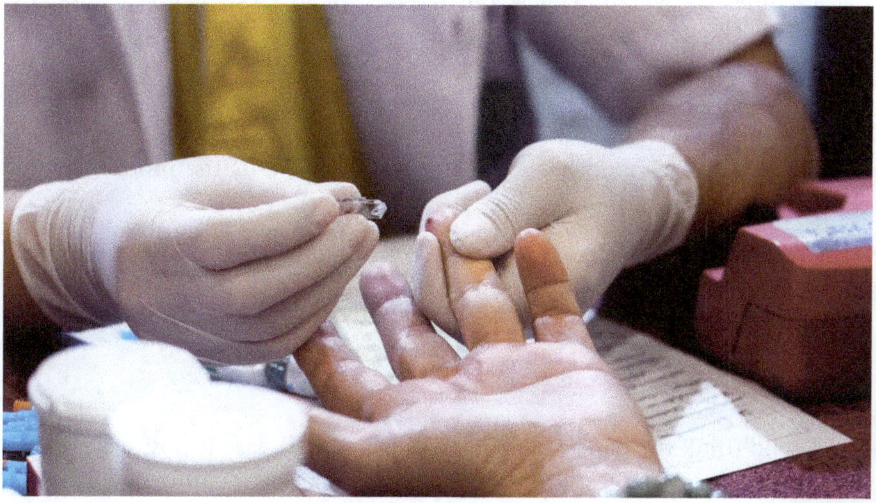

RUCHUDA BOONPLIEN/Shutterstock.com

who lived in the early 1700's. The basic idea of Bayesian analysis is to provide a probability structure to take prior knowledge about a situation and update the prior knowledge based on new information by using conditional probability. There's a formula for this, called Bayes' rule. Sometimes, this leads to surprising results.

Here's an example involving a diagnostic test for a rare disease (like the skin test for tuberculosis, given in many states). The result of the skin test will be either positive (+) or negative (-). A positive result does not mean you have the disease, but only that your chances of having it have increased. A negative result essentially ensures that you don't have the disease. In other words, the diagnostic test updates the original (*prior*) probability that you have or don't have the disease. The diagnostic test is cheap, easy to administer, and has no serious side effects. It's a preliminary test, designed to update probabilities so that those who test positive can then be given a chest X-ray (in the case of TB) to make a final determination. You can analyze the results with conditional probabilities. First the prior (unconditional) probability of having the disease, which is the percentage or proportion of the population with the disease. Suppose that, using empirical data, the prior probability of the disease equals 2%:

P(disease) = 0.02. The conditional probabilities, also found from empirical data, are as follows:

- 95% of people with a certain disease will have a positive reaction ("+") to the diagnostic test. Using conditional probability notation, P("+" | disease) = 0.95
- 8% of people who don't have the disease will also test positive ("false positive").
- P("+"| no disease) = 0.08.
- Prior probability (given above): P(disease) = 0.02

Nancy takes the test and has a positive result. What is the probability that Nancy has the disease?

Why Isn't the Diagnostic Test More Precise?

You would think that since someone who has the disease is positively identified by the diagnostic test 95% of the time and falsely identified ("false positive") only 8% of the time, then if someone gets a "+" on the test, you would be pretty sure they had the disease. This is not the case. The reason is simple: Most of the population doesn't have the disease. In conditional probability terms, P(+ | disease) is different from P(disease | +). This is illustrated by applying Bayes' rule.

BAYES RULE

This procedure updates prior information based on new information. In order for this to be accurate, the prior probability and other relevant probabilities should be reasonably accurate (In other words, you can't just make stuff up). Bayes' rule is a clever use of the multiplication rule for conditional probability. It states that for any events A and B,

P(A and B) = P(A | B) * P(B), from which you get, by rearranging the terms,

P(A | B) = P(A and B) / P(B) = (P(B | A) * P(A)) / P(B)

You flip the conditional probabilities from P(A|B) to P(B|A).

In the case of the diagnostic test, let

A = disease, B = "+" reaction, so the formula reads

P(disease | "+") = P("+" | disease) * P(disease) / P("+")

To solve this equation, we know that P("+" | disease) = 0.95, P("+" | no disease) = 0.08, and P(disease) = 0.02. We need to find P("+"), which can be expressed as:

P("+") = P("+" and disease) + P("+" and no disease) =

P("+" | disease) * P(disease) + P ("+" | no disease) * P(no disease), which gives

P("+") = .95 * .02 + .08 * .98 = 0.0974

And so, P(disease | "+") = .95 * .02 / .0974 = 0.195

Result of the Test

P(Nancy has disease | "+") = 0.195, or 19.5% if you like to use percentages.

The *prior* probability, 0.02, the fraction in the population who have the disease, has been updated to the *posterior* probability, 0.195, where posterior means "given the result of the diagnostic test" or P(disease | "+"). As I mentioned, this may seem low, but remember that only 2% of the population have the disease. Nancy's probability of having the disease is now 19.5%, updated using conditional probability according to her positive test result. Although this posterior probability is small, it's almost 10 times larger than the original probability of 2%. On the other hand, P(no disease | "+") = 1 - P(disease | "+") = 81.5%. It seems paradoxical that the probability of no disease given "+" is higher than P(disease | "+"), but that's because the prior probability of disease is so low (0.02). As we shall see in the next example, this phenomenon is sometimes called "resemblance" stereotyping.

Also, P(no disease | negative) = 0.999, or 99.9%.

These results are typical of a diagnostic test, like the TB skin test, an inexpensive test with few side effects. You can safely eliminate everyone with a negative reaction. Those with a positive reaction don't necessarily have the disease (only a 19.5% chance in this case), so they will get a more comprehensive test, like a chest X-ray, to make a final determination.

Ryzhkov Oleksandr/Shutterstock.com

Lipik Stock Media/Shutterstock.com

Stereotyping - Librarian or Farmer?

On p. 6 of his best-selling book, "Thinking, Fast and Slow," the late Nobel prize winner Daniel Kahneman gives an interesting example of how people can be fooled by "resemblance" stereotyping. Here is the set-up:

> "An individual has been described by a neighbor as follows: 'Steve is very shy and withdrawn, invariably helpful but with little interest in people or in the world of reality. A meek and tidy soul, he has a need for order and structure and a passion for detail.' Is Steve more likely to be a librarian or a farmer?"

When Kahneman presented this description in one of his well-known experiments, most respondents said that Steve was more likely to be a librarian than a farmer because of stereotypical "resemblance," in which librarians seemed to fit Kahneman's description better than farmers. As Kahneman puts it:

> "The resemblance of Steve's personality to that of a stereotypical librarian strikes everyone immediately, but equally relevant statistical considerations are almost always ignored. Did it occur to you that there are more than 20 male farmers for each male librarian in the United States? Because there are so many more farmers, it is almost certain that more "meek and tidy" souls will be found on tractors than at library information desks. However, we found that participants in our experiments ignored the relevant statistical facts and relied exclusively on resemblance. We proposed that they used resemblance as a simplifying heuristic (roughly, a rule of thumb) to make a difficult judgment. The reliance on the heuristic caused predictable biases (systematic errors) in their predictions."

The respondents in the experiment ignored the fact that the population contained 20 times more farmers than librarians. Let's call Kahneman's description of Steve "Shy". Even if P(Shy | Librarian) is greater than P(Shy | Farmer), the preponderance of farmers in the population could make P(Farmer | Shy) greater than P(Librarian | Shy). In other words, there may be more shy farmers out there than shy librarians simply because there are more farmers. This is reflected in the prior distribution of farmers and librarians, which is separate from any stereotyping of Steve. (We only consider farmers and librarians and exclude the rest of the population.)

To compare P(Librarian | Shy) to P(Farmer | Shy) by using Bayes Rule you need to estimate P(Shy | Librarian) and P(Shy | Farmer). In order to identify stereotyping of the type Kahneman described, you could obtain estimated probabilities from an experiment like Kahneman's. You could then use these subjective probabilities to find P(librarian | shy) and P(Farmer | Shy) to demonstrate the importance of prior probabilities in making such determinations.

Here are the conditional probabilities based on a hypothetical experiment of the type Kahneman describes: P(Shy | Librarian) = 0.6, P(Shy | Farmer) = 0.2.

> Librarians are three times more likely to be shy than farmers, however, as Kahneman points out, there are 20 times more farmers in the population than librarians. You can now use Bayes' rule as follows:
>
> P(Librarian | Farmer or Librarian) = 1/21 = 0.047, P(Farmer | Farmer or Librarian) = 20/21
>
> P(Shy) = P(Shy | Librarian)*P(Librarian) + P(Shy | Farmer)* P(-Farmer) = .6 * 1/21 + .2 * 20/21 = 0.22, so
>
> P(Librarian | Shy) = P(Shy | Librarian) * P(Librarian) / P(Shy) = .6*(1/21)/.22 = 0.13, and
>
> P(farmer | Shy) = 1- P(Librarian | Shy) = 0.87

Since there are 20 times more farmers than librarians, if you see a shy person and have to decide if that person is a librarian or a farmer, even though the chance of being shy given that you're a librarian is greater than the chance of being shy given that you're a farmer, your best guess, due to the preponderance of farmers in the population, is that the person is a farmer. In this case, the prior probabilities overwhelm the stereotype.

Bottom Line

Resemblance stereotyping can lead to misclassifying someone because underlying data isn't taken into account. Same for taking a diagnostic test for a rare disease. Bayes' Rule is a way to clarify this.

CHAPTER 17

THE PROSECUTOR'S FALLACY AND SIMPSON'S PARADOX

The Prosecutor's Fallacy

The Prosecutor's Fallacy occurs when a prosecuting attorney in a trial exaggerates or misinterprets probabilistic evidence in an attempt to gain a conviction. A classic example of the prosecutor's fallacy is People v. Collins, in which a couple suspected of committing a robbery in San Pedro, CA in 1968 were convicted based on the testimony of an expert witness brought by the prosecution.

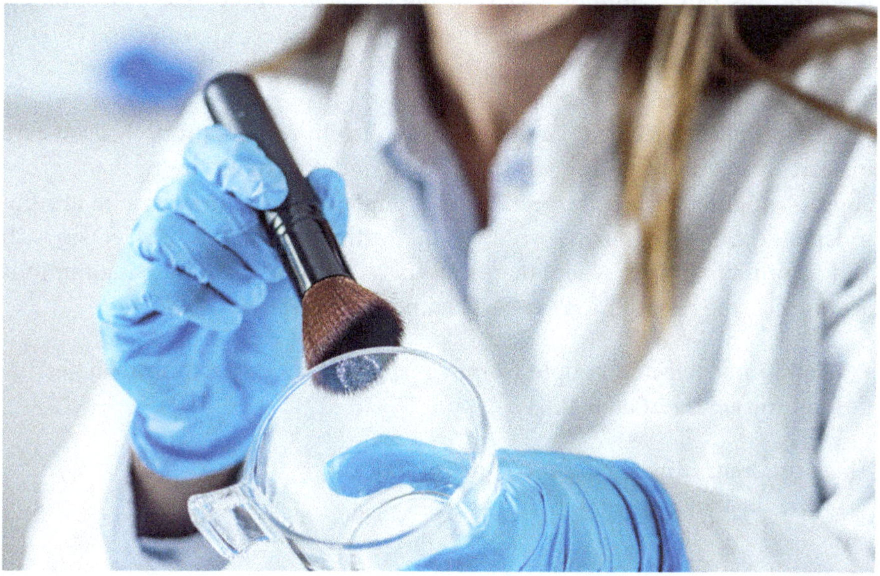

Microgen/Shutterstock.com

In this case the expert witness, a math instructor at a nearby college, introduced the following set of characteristics, based on eyewitness descriptions, along with probability estimates, with no sources. For example, an eyewitness said she saw a white woman with a pony tail at the scene of the crime and that she got in a yellow car driven by a Black man with a beard.

	People v. Collins
Black man with beard	0.1
man with mustache	0.25
white woman with pony tail	0.1
white woman with blonde hair	0.33333
yellow car	0.1
interracial couple in car	0.001
probability estimate (multiplied)	0.000000083
1 in	12,048,193

In addition to having no explanation for the probability estimates, the expert witness proceeded to assume that the characteristics in question were independent events (they're not) and multiplied the probabilities together to get a probability of about 1 in 12 million that a randomly selected person from the population in question (greater Los Angeles area?) would have the listed characteristics. From this, the expert inferred that the chance the suspects were *innocent* was 1 in 12 million. Despite big problems in the expert's testimony, the defendants were found guilty. The defense attorneys appealed the conviction. The appeal reached the California Supreme Court.

If the multiplication rule for conditional probabilities were properly used here instead of the multiplication rule for independent events the relevant probabilities might be much larger. (For example, beard and mustache are not independent and neither are white woman with blonde hair). But let's go with 1 in 12 million and see where that takes us. Assuming that other assumptions are correct, the prosecution's expert claimed to show that:

P(a randomly selected couple from the appropriate population matches the description) = 1/12,000,000.

That's different from saying P(couple is innocent) = 1/12,000,000.

This brings us to *opportunity*. We can ask this question: Given that we have a couple who matches the description, what's the chance that there is at least one other couple who also matches the description? This is the question asked by the statistical experts hired by the defense attorneys who appealed the guilty verdict. Here's how to say this using conditional probabilities:

P(at least two matches | at least 1 match) = P(at least two matches) / P(at least one match).

This is because (at least two matches and at least one match) = (at least two matches).

To find these probabilities using the unsubstantiated estimates provided by the prosecution's expert, we need to know the appropriate population size. When the crime took place in 1968 in the San Pedro area of Los Angeles, the population of the greater Los Angeles area was around 8,000,000 people. We will use 8 million in the formula. This is a rough estimate since some of the population could not have committed the crime. On the other hand, a couple from outside the area could have come in to commit the crime, as sometimes happens in large metropolitan areas. Here are the relevant probabilities and computations, using the expert's number with P(match) = 1/12,000,000.

P(no matches in 8 million people) = (1- P(match))^8,000,000 = 0.5134

P(at least one match) = 1- P(no matches) = 0.4866

P(at least two matches) = P(at least one match) – P(exactly one match) = 0.1443

P(particular match and no other match) = P(match)*P(no other match) = 0.000000043

P(exactly one match) = 8,000,000*P(a particular match and no other match) = 0.3423

Finally,

P(at least two matches | at least 1 match) =

P(at least two matches) / P(at least one match) = 0.1443 / 0.4866 = 0.2966.

Given that there is a couple that matches the characteristics given by the expert, there is almost a 30% chance that there is an additional couple who match these characteristics. In other words, based only on the evidence at hand and accepting the questionable numbers provided by the prosecution's expert, guilt is clearly not established beyond a reasonable doubt.

The improper use of statistics, such as no verifiable probability estimates of the characteristics of the guilty couple, along with improper use of the multiplication rule for independent events, failure to use conditional probabilities, and, most damaging, the misinterpretation of P(match) to mean P(innocent) were emphasized by the defense attorneys. The California Supreme Court overturned the convictions.

Simpson's Paradox

Ground Picture/Shutterstock.com

There are two different drugs used to treat mental illness, a strong drug and a mild drug. Two conditions that are treated by these drugs are depression and psychosis. A researcher looking at data on the effectiveness of these drugs notices that the strong drug has a higher cure rate than the mild drug for each of the two mental illnesses being treated. This makes sense to the researcher because the strong drug is supposed to be more effective than the mild drug regardless of the illness. The reason the strong drug is not always given is because of expense and side effects. When the researcher aggregates the data, she finds, to her amazement, that overall, the mild drug is more effective, percentagewise, than the strong drug. How can it be that a drug is better in each case but worse overall? Here's a two way table with summaries.

	strong drug	mild drug
depression	83/90	250/300
cure rate	92%	83%
psychosis	190/270	58/90
cure rate	70%	64%
both	273/360	308/390
cure rate	76%	79%

The reason for the apparent paradox is that the mild drug is used predominantly for depression, the lesser of the two problems, while the strong drug is used for the more serious problem of psychosis. Although the strong drug outperforms the mild drug in each case, the mild drug overwhelms the aggregate numbers because of its use on the easier malady. This shows how conditional results based on illness type may not hold up in the overall percentages.

Here's another example. The following table shows admission rates to the English department and the P.E. department at a major university, broken down by gender (This is a simplification of a well-known study at the University of California at Berkeley that analyzed admission rates for graduate departments.) For this data, although men do better than women overall, women have higher admission rates than men for each department. The reason for the discrepancy is that more men apply to P.E., the easier department to get into, whereas more women apply to the English department, where it's harder to gain admission. Overall, the number of men who apply to the easier department overwhelms the individual rates.

	women	men
P.E.	55/60	310/350
admission rate	92%	89%
English	280/380	50/75
admission rate	74%	67%
both	335/440	360/425
cure rate	76%	85%

Bottom Line

The prosecutor's fallacy shows how evidence that "matches" a suspect with the crime scene can be misleading. Although it may be very unlikely to find a "match" between evidence found at a crime scene and a suspect, it doesn't follow that the suspect is guilty. Due to *opportunity*, there may be additional matches, providing an element of doubt if only the "match" percentages are used.

CHAPTER 18

THE NOWHERE MAN

Sometimes Johnson randomly walks around town, looking for a nice coffee shop where he can sip a cappuccino, check his email, and read the news on his iPad. One balmy Thursday, Johnson found a pleasant café with a nice outdoor seating area that he hadn't been to before. As he sipped his cappuccino, Johnson took in the beauty of his surroundings and decided to estimate the probability that, starting from his house, he would end up at this coffee shop on such a lovely day. It was summer, and the chance of being sunny and warm was 50%. When it was sunny, the chance that Johnson would take

Mariia Ploshikhina/Shutterstock.com

a walk was 20%. Given that Johnson took a walk, the chance of going to the north side of town was 10%. Given that he went to the north side of town, the chance of not going to the grocery store was 40%. If he didn't go to the grocery store, he would walk through the park where the chance of feeding the koi and then turning left instead of right was 5%. Given that he turned left instead of right, the chance of stopping at a nice older coffee shop with no outdoor seating was 85%. Given that he didn't stop at the older coffee shop, the chance of stopping at this particular coffee shop was 10%. Using the multiplication rule for conditional probabilities and based on Johnson's estimated probabilities (which are accurate enough for the situation), Johnson multiplied to get the chance of being at this coffee shop:

P(being at this coffee shop) = 0.5 * 0.2 * 0.1 * 0.4 * 0.05 * 0.15 * 0.1 = 0.000003.

There was only one problem: Johnson thought of other factors that influenced his decision. The chance of not being affected by allergies was 20%. The chance of not being sick was 75%. Chance of not getting hit by a car while crossing the street was 99%. Multiplying these factors reduced the chances of being at this coffee shop to less than 1 in 2 million.

Upon further consideration, Johnson realized that if he was going to accurately estimate his chance of being at this coffee shop, he needed to include probabilities from the past. For example, the chance of coming to Berkeley to go to college was 1%, and the chance of living within walking distance of the coffee shop was 3%. This reduced his chance of being there to less than 1 in 7 billion. The more Johnson thought about it, the more factors he realized that he could include that played a role in his chance of being at that coffee shop on that day.

The factors were almost endless. Whenever he thought he had the correct probability, he could think of a new factor whose probability he could multiply into the previous result and reduce the overall probability of being there. It seemed like everything he did had some small effect. Since he could always think of a new factor, the only number that was smaller than this continuing product of probabilities, regardless of the origin of the probabilities, was zero! In fact, the only condition needed to end up with a resulting probability equal to zero was a never-ending sequence of probabilities, all fractions less than 1, regardless of their accuracy! This led Johnson to the following conclusion. The overall chance that he would end up at this nice Berkeley cafe, enjoying a cappuccino on this warm, summer day, equaled zero! For the same reasons, the chance was zero that Johnson would be *anywhere*! He was the Nowhere Man. But here he was!

Even if the chance wasn't quite zero, it was so small that being there was like winning the lottery. Except there's a difference between landing at the coffee shop and winning the lottery. To win the lottery, you need a specific configuration of winning numbers. Johnson didn't start his random path with a prize in mind that would require a specific sequence of random choices. He took his walk knowing he would end up *somewhere.* Which leads to this conclusion: If you don't care where you're going, the chance you'll reach your destination is 100%. Even though the chance of randomly going to any particular place is close to zero.

Bottom Line

If you multiply an increasingly long sequence of probabilities, the result approaches 0 no matter what the probabilities are.

CHAPTER 19

THE BLACK SWAN GAME

In his iconic book, "The Black Swan," Nassim Nicholas Taleb discusses unlikely, impactful events, like stock market crashes, which he calls "black swans", events that few predict will happen and that are game-changing when they do happen. Black swans have "fat tails," that is, they seem to occur more often than one might expect. Metaphorically speaking, according to Taleb, black swans and fat tails inhabit a world called "extremistan" where extreme events are more likely to happen than in the everyday world of "mediocristan," where the bell-shaped, normal curve is the probability model for just about everything. Black swans are difficult to predict because, unlike getting "7" when you roll a pair of dice, they seldom happen. (Actual black swans appeared in Australia when none had been seen before. Genetics explains their existence.)

Here are two reasons for fat tails: 1) using the wrong probability models and 2) underestimating the occurrence of rare events by not recognizing the "opportunity" for these events to happen.

The Black Swan Game

Here's a game illustrating 1) and showing why black swans are sometimes not predicted because of improperly using the normal curve (Chapter 24 is about the normal curve). A box contains 10,000 balls: 9,999 are labeled "winner" and one is labeled "black swan". There is one drawing each day. A ticket costs $1. You can buy as many tickets as you want. A ticket doesn't specify numbers like a Mega Millions ticket. It just qualifies you for whatever prize or penalty results from the drawing. After tickets are purchased, the master of ceremonies draws a ball at random from the box. If the selected ball is a "winner," every ticket is paid \$500. If the selected ball is the black swan, every ticket loses \$7 million. The selected ball is then replaced in the box for the next day's drawing. Since there are 10,000 balls in the box, the chance that the black swan is picked equals 1/10,000 = 0.0001 and the chance that a winner is picked equals 9,999/10,000 = 0.9999. To find the expected value, E, for one play of the game (neglecting the \$1 price of a ticket), multiply the probabilities by the payoffs and add, to get

$$E = 0.9999*\$500 - 0.0001*\$7,000,000 = -\$200.05.$$

The game has a negative expected value, with an expected loss of \$200.05 per play. The players won't get a black swan very often, an average of once every 27.4 years with daily play. It's tempting to play the game. Unfortunately, on average, the occasional large, \$7 million loss eventually wipes out the small, \$500 wins. The strategy is clear: Quit while you're ahead and never play again. But at what point should you do that? And what if you're unlucky and lose before you have a chance to quit?

If you have "n" independent drawings and only two possible outcomes on each drawing, "winner" or "black swan", since the probabilities are the same from drawing to drawing, then X(n) = the number of black swans in n drawings has a "binomial distribution" with probability p = 0.0001, mean = n*0.0001, and standard deviation = $\sqrt{n*(0.0001)*(0.9999)}$ (I discuss the binomial distribution in Chapter 25).

Suppose someone plays this game 1,000 times, once a day for 2.74 years. What's the probability that the player is a net winner? Since the loss is \$7 million for a black swan, the only way a player can be ahead after 1,000 days is if a winning ticket is picked every day. This probability equals 0.9999^1,000 = 0.9048. In other words, a player has a 90% chance of not seeing a black swan any time in 1,000 days of play and thus winning a total of \$500,000. If this happens and the player keeps playing, however, the black swan will eventually make an appearance and all gains will be nullified.

There can't be *opportunity* in the black swan game, because the result of a drawing is the same for every player, so if one player gets a black swan, *everyone* does. There are not separate results for each player the way there is for an ordinary lottery. A player can purchase many tickets for the same drawing, which is equivalent to making a bigger bet, but a black swan is selected or it isn't.

It's true that in the stock market, different investors have different strategies and different results, however in a "good" market, investors profit on average, and in a rare crash, investors suffer losses on average. That's the point of the black swan game.

Fat Tails and the Central Limit Theorem

According to the central limit theorem, the normal (bell shaped) distribution can be used to approximate binomial probabilities for large sample sizes, even though the normal curve trails off drastically after 3 standard deviations from the mean. This would not be a problem if the constraints of the central limit theorem were followed, but sometimes this isn't the case and the normal curve is improperly used. In order use the normal curve to approximate binomial probabilities, not only does n, the sample size, need to be large, but, as a rule of thumb, n*p, along with n*(1-p) have to both be larger than 5, which is not the case here. For example, in 1,000 trials of the black swan game, 1,000*.0001 = 0.1, which is considerably less than 5.

The probability graph of the black swan game looks nothing like a normal curve. Here is a probability graph of 1,000 plays of the black swan game, truncated at 20 days (probabilities are essentially 0 after n = 20), along with a picture of the normal curve, and a graph illustrating a coin-tossing approximation to the normal curve when n = 36 (n*p = 18, n*(1-p) = 18).

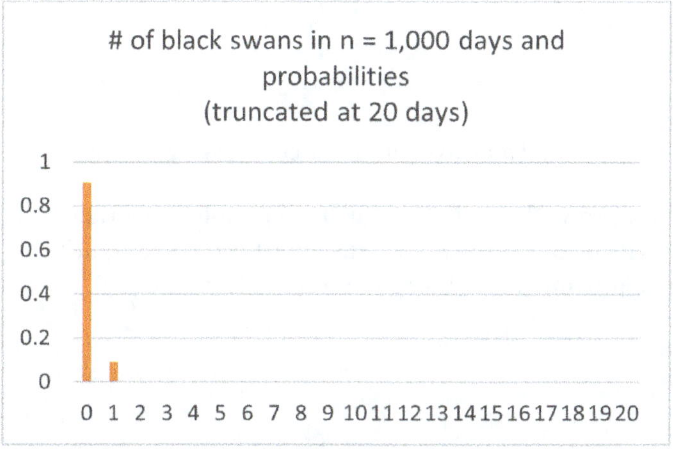

Black Swan Game with n = 1,000 (truncated)

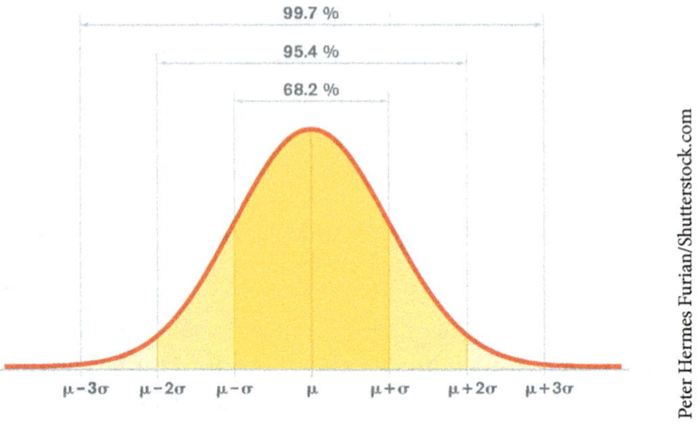

Normal Curve – units of standard deviation from the mean

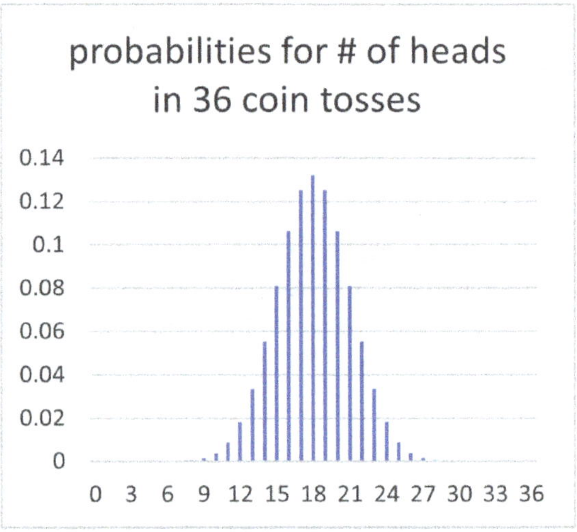

Normal approximation for coin tossing, n = 36

Table 1 below shows that with n = 1,000, even though the right-tail binomial probabilities drop off after $P(X > 0)$, they don't drop off as quickly as the normal curve "approximation." (Using the continuity correction).

TABLE 1 - Black Swan Game (p = 0.0001, n = 1,000)

	binomial	normal
P(X > 0)	0.095	0.1029
P(X > 1)	0.005	0.000005
P(X > 2)	0.0002	0.00000000000002
P(X > 3)	0.000004	0 (to 30 decimal places)

Table 1 is an example of fat tails in the sense that the actual binomial right tail probabilities are much larger than the inaccurate normal approximation. When n = 1,000 days, with the correct, binomial distribution, P(more than 1 black swan) = 0.005 (5 in 1,000), whereas with the normal approximation, P(more than 1 black swan) = 0.000005 (5 in 1 million). With the binomial distribution, P(more than 2 black swans) = 0.0002 (1 in 5,000), whereas with the normal approximation, P(more than 2 black swans) = 0.00000000000002 (1 in 50 trillion) . With the correct, binomial distribution P(more than 3 black swans) = 0.000004, whereas with the normal distribution, P(more than 3 black swans) = 0 rounded off to 30 decimal places.

In the black swan game, the sample size will be large enough to use the normal approximation when n*p > 5, which in this case means n > 50,000, or after 136 years. Since n increases with daily drawings, not with additional players, the normal approximation won't be accurate in our lifetime.

Can the Black Swan game have positive expected value?

We found the expected value for one play of the game (neglecting the $1 price of a ticket): E = 0.9999*$500 - 0.0001*$7,000,000 = -$200.05.

The player has a negative expected value, with an average loss of $200.05 per play.

A way to make this game have positive expected value is to decrease the probability of a black swan. You can solve for the expected value break-even point, as follows:

$$E = (1-p)*500 - p*7{,}000{,}000 = 0, \text{ which yields } p = 0.000071 = 1 \text{ in } 14{,}001.$$

So instead of a box with 10,000 balls, you can add, say, another 5,000 balls, so that there are now 15,000 balls in the box, with 1 marked "black swan" and 14,999 marked "winner." Everything else is the same as before: A ticket marked "winner" pays $500 and a "black swan" causes a $7,000,000 loss. Expected Value is now

$$E = (14{,}999/15000)*500 - (1/15{,}000)*7{,}000{,}000 = \$33.30.$$

The game has positive expected value and, even though the dynamics are similar to the initial game, with prudent betting, the occasional, large losses won't overwhelm the many small wins, and the investor (gambler) will make a long run profit.

Bottom Line

The Black Swan game shows how a common misuse of the normal curve can provide misleading probabilities of a black swan and create a situation for fat tails.

CHAPTER 20

WAS HUMPTY DUMPTY ACCIDENT PRONE?

Was Humpty Dumpty accident prone? Did he have any previous accidents before his fall? Did he have a predisposition to have accidents (Why would he sit on a wall)? Was he drinking alcohol or on drugs? (Again, why would he sit on a wall?) What does it mean to be accident prone?

HitToon/Shutterstock.com

According to the Oxford Languages dictionary, "accident prone" means "tending to be involved in a greater than average number of accidents." Having "a greater than average number of accidents?" Compared to whom? Let's compare Humpty Dumpty's accidents to those who have the opportunity for similar accidents, namely, other nursery rhyme characters. Even though a spider sat down beside her and scared her away, there's no evidence that Little Miss Muffet fell off the tuffet or had any other accidents. There's no record of any accidents for Mary, even though she was quite contrary. While it's true that Jack Sprat could eat no fat and his wife could eat no lean, there's no evidence that either of them had any accidents. Old King Cole was a merry old soul and accident-free as far as we can tell. Hickory Dickory Dock, the mouse ran up the clock, and when the clock struck one, the mouse came down, but there's no evidence that he fell down. The farmer's wife cut off the tails of the three blind mice, but I wouldn't call that an accident. On the other hand, when Jack and Jill went up the hill to fetch a pail of water, Jack fell down and broke his crown and Jill came tumbling after. Accidents for both of them.

To summarize: Humpty Dumpty, and Jack and Jill each had one accident. Little Miss Muffet, Jack Sprat, his wife, Old King Cole, the mouse who ran up the clock, and Mary (quite contrary) didn't have any accidents. That means 3 accidents for 9 characters, an average of 0.33 accidents per character. In other words, according to the Oxford Languages Dictionary definition, Humpty Dumpty, along with Jack and Jill all had more accidents than average (one accident each), and so they were all accident prone. Using nursery rhyme characters in this context is a joke, but so is the Oxford Languages Dictionary definition of accident proneness.

In fact, for any set of measurements, some will be greater than average and some will be less than average. This has nothing to do with accident proneness or anything else except the definition of average. We don't want to misclassify unlucky people as being accident prone. For example, why should someone's insurance rates go up just because they're unlucky?

Another Definition - Predisposition

Wikipedia says, "Accident-proneness is the idea that some people have a greater predisposition than others to suffer accidents, such as car crashes and industrial injuries. It may be used as a reason to deny any insurance on such individuals." Now, we're getting somewhere. What is a "predisposition?" The Oxford languages dictionary says that a predisposition is "a liability or tendency to suffer from a particular condition." An example of a predisposition is a genetic predisposition where "you are likely to inherit a trait from your parents." How do you tell if someone has a predisposition for having accidents? Does someone who gets struck by a car driven by a drunk driver have a predisposition for having accidents? Or is it just bad luck? Is there a predisposition test, like an IQ test? Should we go the eugenics route and try to breed people without predispositions?

Rene Jansa/Shutterstock.com

PERSONALITY TRAITS AND ACCIDENTS

In "Personality Characteristics of the Accident Involved Employee" (Journal of Business and Psychology, Vol. 2, #4, 1988), Curtiss P. Hanssen provides a survey of some of the research in this area. He discusses the notion of the "differential accident liability concept, " which is, in Hanssen's view, an updated version of accident proneness. Hanssen says in his introduction, "The research relating personality traits to industrial and traffic accidents is reviewed...This article will focus on the personality traits that have been associated with accident occurrences over 65 years of research." Hanssen further says, "If personal differences are discovered that are associated with accident occurrences, then future job applicants with those traits may not be hired or will be placed in low risk positions."

What are these personality traits? Hanssen sites a 1963 study which found that "extroverts to have had more traffic accidents and safety violations than introverts." Another study in 1981 presumably corroborated this, finding that "extroversion significantly correlated with both number of accidents and violations." In an industrial setting, another study found that there was "a high positive correlation between number of accidents and extroversion for males but not for females."

In the conclusion section of the article, Hanssen lists eight relationships between "personality related variables and accident occurrences," six of which are "extroversion," "aggression," "social maladjustment," "general neurosis," "specific neurotic conditions such as anxiety and depression," and "impulsivity."

Don't hire extroverts? Or applicants who are anxious or depressed? Maybe it depends on the type of job the applicant is applying for. I don't want to travel in an airplane flown by a depressed pilot. Will my auto insurance company tell me they're raising my rates because I'm neurotic? I can't think of a better way to make me more neurotic than I already am. Hanssen says that "the anxious person struggles to control his internal anxiety and panic, giving less attention to the details of safe performance," and that "the depressed person is preoccupied with his internal world of ruminations, not with the exigencies of the external world." On the other hand, Hanssen says "several studies, however, have found no relationship between accidents and neuroticism." Whew! If you search through enough studies, you'll find the answer you want. He cites another study that "found that extremely neurotic subjects had a higher accident rate than other groups, even though the 'average' neurotic did not." (The average neurotic?) Hanssen then says, "To complicate matters further, two studies have reported an inverse relationship between accidents and neuroticism. That is, neurotic drivers tended to have better safety records than nonneurotic drivers." What do we do with all this? The conclusions of the studies cited by Hanssen are based on correlations between certain groups and accident levels. But correlation does not mean causation. Care must be taken.

Hanssen also reviews numerous studies to try and identify human factors that may be correlated with accidents. As we have seen, a byproduct of reviewing numerous studies are false positives (if you do enough testing you'll find significant results that are really just due to chance), yet Hanssen doesn't mention it. Generally, the research focuses on which subgroups of the population are more likely to have accidents than others. This is useful, given the seriousness of accidents, but scant attention is paid to chance, which appears to be a factor in many accidents. Insurance companies would have a problem if all accidents other than the obvious culprits like drugs, alcohol, texting while driving, etc., were caused by chance.

Statistical Significance

Even if there are personality traits and other factors that are correlated with accidents, there should be a way to separate the accident prone from the unlucky, that is, to identify those who have *significantly* more accidents than average compared to co-workers or peer groups. The way to do this is by using p-values. Chance is not a predisposition. I'm not challenging the predisposition concept, but to be labeled accident prone, someone should have more accidents than by just being unlucky.

One of the challenges of a company identifying whether an employee is accident prone is that there may be no clear procedure for separating bad luck from statistical significance. A few years ago, I had a consulting project in which the client was suing her employer, a large company with many employees, that was trying to fire her because she was, according to the company, *accident prone*. This was a successful company who could afford to hire statisticians to make such determinations. I was the statistician hired by the client's attorney. When we were presented with evidence of accident proneness, the company provided a list of accidents the client had incurred over the past few years. None of the accidents was serious enough to require hospitalization. The worst accident was when my client was walking down steps and strained her back. This required a heating pad and ibuprofen and that was that.

When we asked why this data showed that my client was accident prone, the company statistician assigned to this case said that she had more accidents than average. It was never suggested that she was negligent or careless (no evidence of that). They never mentioned statistical significance and seemed confused when I asked about it. Even if they had calculated the average number of accidents properly, the company apparently didn't realize that some people will have more accidents than average and some will have fewer just by the definition of average, and due only to chance, having nothing specifically to do with predispositions or anything else.

When we asked for data showing that the client had more accidents than average, the company's statistician provided accident data for the company's entire workforce, including people in desk jobs. My client worked outside and was exposed to many more hazards than the desk job people, so, in addition to the fact that there will always be some people who have more accidents than average, the "average" that the company provided was inappropriate.

According to the company's presentation, my client's rate of accidents was only slightly higher than average, even when you included desk jobs. We didn't have to work hard. The case was heard by a mediator, who ruled in favor of my client. She still works for the company at the time of this writing, which is many years after the case was concluded.

A Hypothetical Example

Suppose that in a manufacturing facility, a warehouse worker has had 7 accidents in the past three years, all requiring treatment by the facility's medical personnel. The worker's supervisor accuses the employee of being accident prone, since a company accident database over the previous 3 years shows that of the 375 warehouse and assembly line workers, there were 520 accidents, yielding an average of 520/375 = 1.39 accidents per worker. The company statistician does the above analysis as follows:

520 accidents, 375 workers, 520/375 = 1.39 accidents per worker

Under certain conditions, the Poisson distribution provides probabilities for counts of accident occurrences (don't worry about the mathematical details). Using an average of 1.39 accidents per worker, I found the p-value: $P(X \geq 7) = 0.0006$. The p-value is considerably less than 0.05, indicating that this worker has had a significant number of accidents and is not just unlucky. However, assuming that the company tests all its employees for accident proneness, there is the possibility of false positives due to the opportunity provided by multiple tests (someone whose p-value is significant but is really just unlucky). Similar to a previous discussion about the Mega Millions lottery, it's useful to find the probability that, assuming that nobody is accident prone (innocent until proven guilty) that *someone* has at least 7 accidents. Assuming that every worker has the same chance of having an accident and that accidents are independent from worker to worker, we can compute the probability that *someone* has at least 7 accidents as follows: Letting X = number of accidents for a particular worker, then with 375 workers and $P(X \geq 7) = 0.0006$, and noting that P(a worker has < 7 accidents) = $1 - P(X \geq 7)$, we find that

P(*someone* has at least 7 accidents) = 1- P(*nobody* has at least 7 accidents) =

$$= 1 - (1 - P(X \geq 7))^{375} = 0.20$$

This is not significant but in the range of bad luck. We have switched the p-value from a particular worker to *someone*. More investigation needs to be done for the worker in question, but it's not unlikely that *someone* from this population would have had at least 7 accidents, assuming as I have done, that all workers are equally likely to have accidents.

Run Over by a Bus ...Twice

Here is another hypothetical example that illustrates the challenges in finding appropriate accident data and conducting an analysis for accident proneness. The dictionary definition of accident proneness mentions "car crashes and industrial injuries." The example is hypothetical but I use real bus accident data.

Ellen, a 35 year-old computer programmer, was walking to work on a sunny, pre-pandemic day in 2019 in the Times Square area of New York City where she worked, when she got struck by a bus. Fortunately, she was not seriously injured. Ellen didn't drink or take drugs, wasn't taking any medications that might impair her judgement, wasn't careless, wasn't texting, and was in good health at the time of the accident. Furthermore, she had no history of accidents more serious than stubbing her toe or spilling her soup and no knowledge of any predisposition that would have made her more prone to accidents than anyone else. With just one accident, she could hardly be called accident prone.

A few months after Ellen recovered from her injuries, she was walking to work again, when two blocks from the scene of the first accident, she was struck by *another* bus. Fortunately, Ellen wasn't seriously injured this time either. Her daily routine hadn't changed, and she hadn't developed any bad habits that might have impaired her judgement.

Was being struck by a bus twice in the same year in the same part of town due to chance (bad luck) or was Ellen accident prone, with a significant number of accidents and possibly with a predisposition to have accidents?

To use p-values, we need to estimate two things:

1. The probability of a person having the observed number of accidents just due to chance (bad luck). This will become the p-value.
2. The "opportunity" for such things to happen, measured by the chance that *someone* would have the same rate of accidents.

I'll use past data to estimate the probability of being struck by a bus in the Times Square area of New York City, but the denominator of the probability fraction won't be the population of the United States the way it is for the probability of being struck by lightning or being killed by a hornet, wasp, or bee sting. Instead, I will use the number of pedestrians in this area who have the *opportunity* to get struck by a bus, similar to the number of people who buy tickets to a lottery, in this case, a bad luck lottery.

Bus Accidents in New York City

evgenii mitroshin/Shutterstock.com

For those interested in U.S. vehicle accident rates, the NHTSA (National Highway Traffic Safety Administration) maintains extensive data, particularly for accidents with fatalities. They have an online query tool that enables a researcher like you or me to build custom queries for specific accident data.

Using the NHTSA query tool, I found bus accident data. In 2019, there were 14,000 bus accidents nationwide involving a non-fatal injury. Let's assume, hypothetically, that about 10% of these accidents, or 1,400, were in New York City where Ellen lives (NHTSA only has *fatal* accident data broken down by state). According to the website, times square nyc , the Times Square area had a pre-pandemic average of 360,000 visitors per day, spread out evenly over the days of the week, which is where Ellen had her accidents. According to another section of the same website, about 172,000 of these 360,000 pedestrians were office workers like Ellen, rather than tourists, so they would be around on a daily basis and thus have the *opportunity*, like Ellen, to have multiple bus accidents in the Times Square area.

If there are 1,400 bus accidents in New York City annually that result in non-fatal injuries, then, there are an average of about 4 accidents per day. Since

Times Square has the heaviest pedestrian traffic in New York City and since this area also has heavy bus traffic, I'll estimate that an average of about 2 of these 4 daily bus accidents take place in the Times Square area. Subjective probability at its finest, but this is not a casino.

Let's assume that all pedestrians in the Times Square area are entered in a bad luck lottery in which everyone is equally likely to be struck by a bus. As I mentioned, the denominator of the probability fraction is not the population of the U.S., but only those who have the opportunity to be run over by a bus more than once in the Times Square area during a particular time period. We're assuming that chance of being hit by a bus on a given day in the Times Square area equals about 2/360,000. Since 172,000 of these 360,000 pedestrians are office workers like Ellen, about one office worker who works in the area (more precisely, 2*172/360 = 0.96 of an office worker) will get hit by a bus daily, with probability 1/172,000 = 0.000006. What about the probability of having at least 2 such accidents in a year?

Finding the P-Value – Binomial and Poisson Distributions

You can model probabilities for the number of accidents Ellen has in a year with the binomial distribution (discussed in Chapter 25). Since we are looking at office workers like Ellen who go to work on weekdays, there are 5*52 = 260 weekdays in a year. We'll reduce this sample to n = 150 days per year to account for holidays, sickness, vacations, and recovering from previous accidents. Since the probability of an office worker being hit by a bus on a random day equals about 1/172,000 = 0.000006, we will use this as our estimate of Ellen's chance of being struck by a bus on any given day of the 150 eligible days. We also assume that the probabilities are independent and remain the same from day to day. Letting X denote the number of accidents Ellen has in n = 150 days, we use software to find that

$$P(X \geq 2) = 0.0000004, \text{ or about 1 in 2.5 million.}$$

I could also use the *Poisson* distribution to model probabilities for the number of accidents Ellen has in a year. When the chance of an occurrence ("accident") is very low on a given day, with virtually no chance of more than one occurrence on a given day and when occurrences are independent over time, then the Poisson and binomial distributions will provide similar probabilities.

In this case, if the chance of an office worker being hit by a bus on a random day equals 0.000006, then the Poisson "average," $\lambda = 150 * 0.000006 = 0.0009$, and so, still letting X denote the number of times in a year that Ellen is struck by a bus we use software to find: $P(X \geq 2) = 0.0000004$.

This is the same, rounded off to seven decimal places, as the corresponding binomial probability. So it seems that Ellen is accident prone in the sense that her accidents are significant relative to others with the same opportunity. However, this doesn't take into account the opportunity for *someone* to be unlucky. Recall that in the Mega Millions lottery, since every ticket has the same chance of winning the jackpot, the important probability is the chance that *someone* (some ticket) wins the jackpot, not the chance that a particular person wins the jackpot. In this case, like a bad luck lottery, an important probability is the chance that *someone* gets hit by a bus twice in a year, not the probability that *Ellen* gets hit by a bus twice in a year. If we assume that everyone has the same chance of being hit by a bus twice in a year (some people may be more likely than others, but that's not a major consideration here), and that accidents among people are independent, then the chance that *someone* gets hit by a bus twice in a year is computed as follows.

If X = # of times a particular person, like Ellen, gets hit by a bus at least twice, then, since there is opportunity for 172,000 workers to have at least 2 bus accidents, and since $P(X \geq 2) = 0.0000004$ (Ellen's probability), then

P(*someone* gets hit by a bus at least twice) = 1 – P(nobody gets hit by a bus at least twice)

$$= 1 - (1 - P(X \geq 2))^{\wedge}172{,}000 = 1 - (1 - .0000004)^{\wedge}172{,}000 = 0.06649.$$

In other words, given the observed data, the chance that *someone* gets hit by a bus at least twice in the given time period is not significant and can be attributed to bad luck. Ellen just happens to be the someone.

All the data was taken from NHTSA and the Times Square websites except that the estimate of bus accidents in New York was taken from NHTSA bus accidents nationwide, and the estimate of bus accidents in the Times Square area was taken from the estimate of bus accidents in New York. I showed by using available data, that it's reasonable to conclude that *someone* getting struck by a bus at least twice in a year is attributable to luck. Although alternatives like accident proneness or predispositions may exist, further study of the causes of Ellen's (or anyone's) accidents would be called for.

The Importance of Opportunity

It's important to realize the role of opportunity in deciding whether what we have observed is due to bad luck (chance) or to something else. Just as the sale of millions of Mega Millions lottery tickets provides the opportunity for someone to be a lucky jackpot winner, the number of bus accidents in a particular area, along with the flow of pedestrians, provides the opportunity for *someone* to get run over by a bus at least twice in a year, just due to chance. It's sometimes hard to realize the importance of opportunity, which, in this case, is provided by pedestrians who *didn't* get struck by a bus. People are interested when they see a media story about someone who got struck by a bus and even more interested if someone got struck by a bus twice. But you never see a story about someone who *didn't* get struck by a bus. For one thing, nobody would read it, except maybe a few statisticians. It may seem odd that the people who didn't get struck by a bus are a necessary part of deciding whether someone who did get struck by a bus was just unlucky, but that's the nature of chance.

A Statistician, an Insurance Agent, and Carl Jung Walk Into a Bar

The famous Swiss psychiatrist, Carl Jung, when writing about *coincidences* that he thought weren't adequately explained by chance, suggested that there might be an unknown force at play. He called this unknown force "synchronicity," and defined it as "a hypothetical factor equal in rank to causality as a principle of explanation."

When discussing the Mega Millions lottery, I mentioned apophenia, "the tendency to perceive a connection between unrelated or random things." "Synchronicity" and "apophenia" are related, as the following example shows. Suppose you spill your corn flakes at breakfast and believe this is an omen foretelling that you will have a serious accident later that day. And suppose that as you're driving to work, you get in an automobile accident. Carl Jung would say that the "omen" wasn't a random thing but instead, because it was too unlikely to be caused by chance and because nobody would seriously argue that spilling your corn flakes would *cause* an accident, it's an example of *synchronicity.* An insurance agent would say that you're accident prone. A statistician would say that you were just *unlucky* and that any explanation that involves spilling your corn flakes is an example of *apophenia*.

Mike Orkin

On the Lookout

Those who are on the lookout for synchronicity or predispositions don't always look at the big picture but perhaps only look at news stories highlighting people who have bizarre, unlikely experiences that seem to be unexplainable by chance. This is misleading. Most news stories don't mention the people who could have had unlikely accidents but didn't, like people who bought lottery tickets and didn't win. You can look at events like getting struck by a bus in a broader framework. Here are related probabilities:

- P(Somebody, somewhere in the world, gets struck by a vehicle at least twice in a year)
- P(Somebody, somewhere in the world, gets struck by a bus at least twice in a year)

Sebastian Gora Photo/Shutterstock.com

A few years ago, I almost got run over multiple times when walking around a city in India, once by a bus, once by a cow, twice by a motorcycle, and once by a tuk-tuk (small taxi). If I had actually been run over, it would have been bad luck (I don't think I'm accident prone, but who knows?), and maybe I would have made the news. When I crossed the street, I felt like I was in the video game Frogger. I was determined to win, which I did in the sense that I got to the other side of the street unharmed, although my companion, who didn't venture into the street, nearly had a stroke. There were no news stories about me.

Bottom Line

There may be different ways to define accident proneness, but identifying someone as having more than an average number of accidents is not enough. Somewhere in the process, there should be a test for statistical significance.

CHAPTER 21

CORRELATION

"In the beginning, everything was linear. Then, things leveled off."
– Anonymous statistician

OMENS AND DATA

Looking for relationships and correlations is part of the human condition. Our ancestors learned that if there was a rustling in the tall grass, they were about to be attacked by a lion. Or maybe it was just random rustling, like the wind. Better to play it safe and leave the area. Dark clouds indicated a storm was coming. Shrieking meant monkeys. A hoot meant an owl. When the earth shook, it meant that an elephant, or maybe a rhinoceros, was headed your way. Or maybe it was an earthquake. These were correlations. What should you do, if anything, when there was a full moon? Should you harvest your crops? Make a sacrifice? What about dead fish washing up on the beach? Was this an omen? Maybe the sea was angry that day. What about when the stars are in a certain alignment? What should you do?

Webster's Merriam dictionary says that an omen is "a happening believed to be a sign or warning of some future event." By this definition, correlated data is the modern version of an omen. Software and statistical analyses allow us to decide if there are meaningful correlations and make predictions if that is warranted.

The *correlation coefficient* measures the linear relationship between two variables. For correlated variables, *regression* provides an optimal (in a sense) way of predicting one variable from another when there is a linear relationship between them.

Square Footage vs Selling Price – A Linear Relationship

ImageFlow/Shutterstock.com

A study is done to see if there's a relationship between the size of a house, expressed by square footage in thousands of square feet, and selling price, in millions of dollars. The "bivariate" data, consisting of pairs of numbers is collected and put in a table of X, Y pairs, in Excel or Google Sheets or elsewhere, where X = square footage and Y = selling price, as shown below. The data in the table can then be put into a graph, called a "scatter plot", as shown below.

The scatter plot is a useful visual tool. You can quickly see patterns that aren't obvious from the data table. For example, in the square footage vs selling price scatter plot, you can see that there is a "positive correlation" between X (square footage) and Y (selling price): As square footage increases, so does selling price, in a linear way.

Square ft.	$ price
1	0.6
1.5	0.5
2	1.1
2	2.1
2.5	1.3
2.8	2.3
3	1.7
3.2	2.5

Square footage, X, is called the "explanatory" variable because it helps explain selling price. Selling price, Y, is called the "response" variable because it is responding to square footage. This does not mean that square footage causes selling price, only that the two variables are related. Because of the relationship, we can use a linear formula, based on the scatter plot, to predict selling price (response variable) based on square footage (explanatory variable). In the present scatter plot, points slope up in a way that follows the trajectory of a straight line. Even though the points aren't all on a straight line, an appropriate straight line might be a good way to predict Y based on X. In addition to ever-present random variation, there are other explanatory factors, so for studies like this, the points on the scatter plot are not all on a straight line, but drift upward in a linear way. In the case of square footage vs. selling price, location, condition, and age also help explain a house's selling price.

Predicting the Weather

Chokniti-Studio/Shutterstock.com

A weather forecaster who is also a degenerate gambler decides that a good way to predict the next day's temperature is to roll a die and predict warmer weather as the number on the die increases, according to the following formula (it's summer in Berkeley, CA):

Aliaksander Karankevich/Shutterstock.com

die roll	Prediction
1	70
2	75
3	80
4	85
5	90
6	95

The forecaster does this for a week with the following results, where X is the die roll prediction and Y is the actual high temperature the next day.

die roll	X prediction	Y temperature
1	70	82
5	90	78
3	80	93
4	85	68
2	75	89
6	95	90
4	85	88

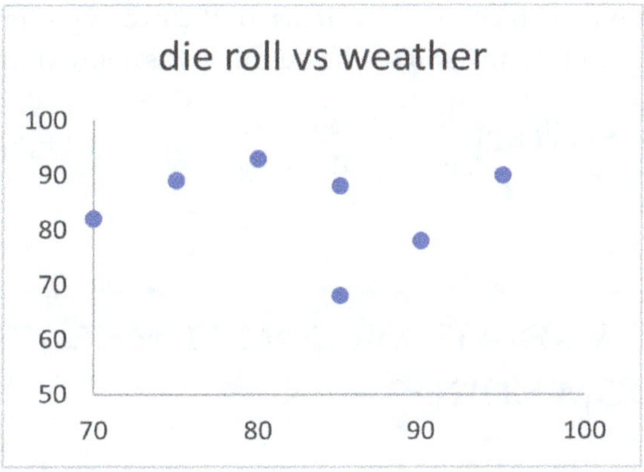

As you can see from the scatter plot, there is no apparent correlation, either positive or negative. It seems like the best straight line to fit the data is a horizontal line, which means that the result of a die roll is useless for predicting the weather.

The Sample Correlation Coefficient

The scatter plot is a useful visual tool to help identify whether correlation exists, but in addition to a visual tool, you need a formula to measure correlation. The formula that does this is called the sample correlation coefficient because it's based on sample data, and is denoted by the letter r. The sample correlation coefficient has these properties:

- $-1 \leq r \leq 1$ (r is between -1 and 1)
- If r = 1, there is perfect positive correlation. All the points on the scatter plot lie on a straight line that has a positive slope
- If r = -1, there is perfect negative correlation. All the points on the scatter plot lie on a straight line that has a negative slope.
- If r = 0, X and Y are uncorrelated; there is no correlation between X and Y.
- r close to 1: strong positive correlation.
- r close to -1: strong negative correlation.
- r close to 0: little correlation

Here are the sample correlation coefficients for the two hypothetical examples. Look at the scatter plots first and see if you can guess approximate values for r.

- square feet vs selling price: r = 0.82
- Dice roll vs weather prediction: r = - 0.08

Random Variation, Sometimes Known as Random Error

You might wonder why the die roll vs weather example has a slightly negative correlation coefficient. How could there be any correlation at all between rolling a die and tomorrow's weather, you might ask? You would be right. There is no correlation between the roll of a die and tomorrow's weather. The reason that r is not exactly 0 is because of random variation. There is usually enough randomness due to sampling to make r slightly positive or slightly negative, even for uncorrelated data. There are statistical methods to decide whether there is actual correlation or whether a non-zero r is due to chance

Correlation vs. Causation — Polio Vaccine vs. Soft Drink Sales

Most data involving correlation is *observational.* The data is obtained by *observing* a process rather than as the result of an experiment in which you can manipulate variables to reduce or eliminate the effects of confounding background variables. As a result, even if there's a strong correlation between two variables, it's not cool to make statements about cause and effect. There might be other factors behind the scenes that are causing things.

In the 1950s, when the polio vaccine was being developed, researchers were trying to better understand how the polio virus worked. Someone noticed that there was a positive correlation between polio incidence and soft drink sales. When soft drink sales increased, so did polio incidence. A researcher concluded that "soft drinks cause polio." While it was true that there was a positive correlation, and while it was true that soft drink sales *predict* polio incidence, causation is another story. It turned out that there was a background variable that was responsible for the correlation: the weather. In warm weather, more soft drinks are consumed. The polio virus is more active in warm weather, so polio incidence is higher. There is no causal link between soft drink consumption and polio. One doesn't cause the other. And so it goes.

Bottom Line

The correlation coefficient measures the linear relationship between two variables, the "explanatory" variable and the "response" variable. For observational data, correlation doesn't mean causation. There may be background variables that are driving the process.

CHAPTER 22

REGRESSION

"Prediction is very difficult, especially if it's about the future"
– Niels Bohr, Nobel Prize winning physicist. Or maybe it was Yogi Berra.

Regression to the Mean

When the 19th century statistician, Francis Galton, was studying genetics and the correlations between hereditary variables, he discovered a phenomenon that he called "regression to the mean," which he discussed in a research article "*Regression towards mediocrity in hereditary stature*." Galton's discovery, which has been used and misused in many situations, is as follows: Suppose you have a random sample of fathers. Suppose you take just the tallest fathers and then find the average heights of their adult sons. Since you started with a random sample to begin with, this average tends to the average of all the fathers, not just the tallest ones, and has thus "regressed" to the mean. The reason is this: The tallest fathers are not a random sample but instead selected because they are the tallest. Even if there were a genetic connection for heights, the sons' heights would tend to "regress" to the average height of the entire group, which includes the tallest fathers. The "mediocrity" in the title of Galton's paper was probably intended to attract readers. This is the origin of the term "regression" used for linear prediction.

The Regression Line

If there is a linear correlation between X and Y, you can predict Y based on X by using a straight line, which has the formula $Y = a + b*X$. There are an infinite number of straight lines. You need to find the best one to fit the data points in the scatterplot. This line is called the "regression line," or "least squares line."

To find the regression line you consider the square root of the average squared distance from a prospective line to each of the points on the scatter plot (sometimes called the RMS, or "root mean square") and then use the line for which the RMS is the smallest. The RMS of a straight line essentially gives the average distance points on the scatter plot are from the line. The distance from a line to a point on the scatter plot is sometimes called "error" because it measures how far off the line is from the point. It's impossible to find the regression line by doing a line-by-line computation because there are too many lines under consideration, but you can do it using minimization methods in algebra or calculus. You can then use this line to predict Y for a given value of X.

Below is the scatter plot for square footage vs selling price, with the regression line and its formula on the graph. For example, if $x = 3$, then, using the regression line, you would predict $y = 0.8125*3 - 0.3156 = 2.12$

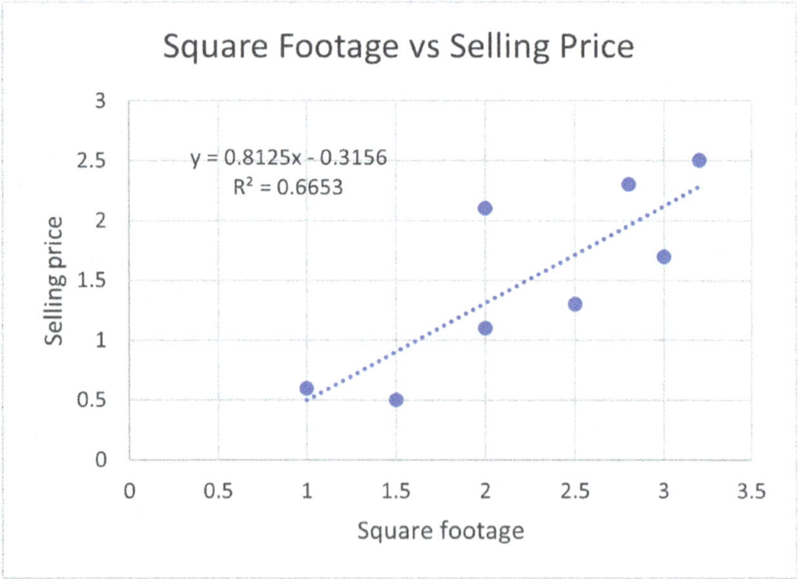

The Coefficient of Determination: R^2 (or r^2)

If the data points on a scatter plot all fall on the regression line, there's perfect correlation: r = 1 or r = -1. If a point (x, y) is on the scatter plot the regression line perfectly predicts y from x. There is no error. If there isn't perfect correlation (there never is with sample data), that is, if the correlation coefficient is not equal to 1 or -1, there is at least one point on the scatter plot that is not on the regression line. For a point, (x, y) that is not on the regression line, the line doesn't predict y exactly for the corresponding x. Instead, there is "error" in the prediction, measured by the vertical distance from y to the regression line.

It turns out that r^2, the square of r, tells you the percentage of the variation in Y that is "explained" by the regression line. For example, if r = 0.8, then $r^2 = 0.64$. This means that 64% of the variation in Y is explained by the regression line. If r = 1 or r = -1, then $r^2 = 1$. All the points on the scatterplot lie on the regression line, so there is no error in predicting Y for the points on the scatterplot when using the line. In other words, if $r^2 = 1$, 100% of the variation in Y is explained by the regression line.

For square footage vs selling price, r = 0.82, so $r^2 = 0.67$. This means that 67% of the variation in selling price is explained by the regression line based on square footage vs selling price. The rest of the variation is presumably explained by other relevant variables, such as location, condition, and age of the house, along with random variation.

If r = 0, then there is no correlation and the coefficient of determination equals 0. In this case, none of the variation in Y is explained by the regression line. In the die roll vs weather example, r = -0.08, so $r^2 = 0.0064$, which means that when you account for random variation, none of the variation in weather is explained by the roll of a die.

Berkson's Paradox – Bacon and Eggs

Nancy likes to go to restaurants and get bacon and eggs. Because of this, she only goes to restaurants that have either good bacon or good eggs or both. She

Alexander Prokopenko/Shutterstock.com

looks at restaurant data, including the restaurants she goes to, and sees, based on a rating system of 1 to 10 that there is a positive correlation: Restaurants with good bacon tend to also have good eggs. Nancy looks only at the restaurants she goes to and discovers something weird: Instead of a positive correlation between bacon and eggs, there is a *negative* correlation. What's wrong?

Here's the data table, scatter plot, and regression line for Nancy's restaurant data:

bacon	eggs
2	8
5	2
5	5
6	3
6	8
9	4

r = -0.42

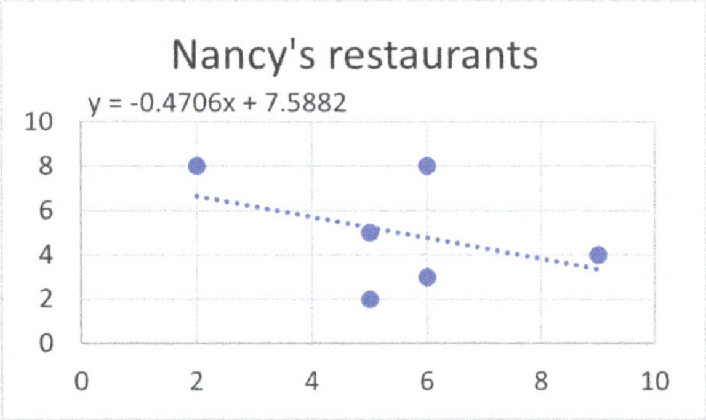

The problem is that there is "selection bias" in Nancy's sample: She's leaving out restaurants that have both bad bacon and bad eggs. In other words, she's leaving out the lower area of the scatter plot. If you include this area, you get the original data, in which there is a positive correlation between the quality of bacon and eggs. This is called "Berkson's Paradox."

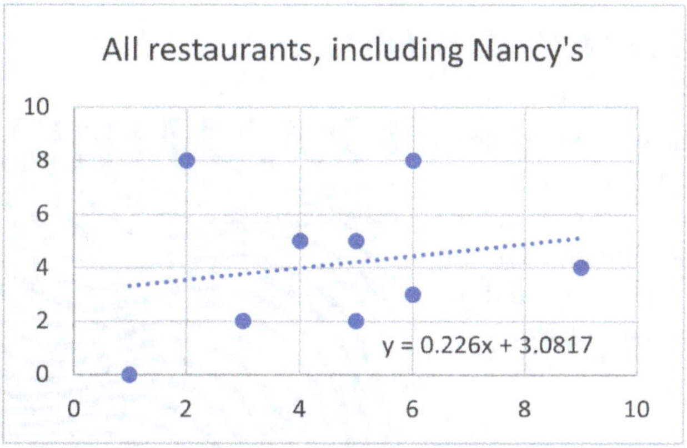

PREDICTING THE PRICE OF ETHEREUM

In a 2021 Reddit article titled, "Modest Prediction of ETH Price = $6400 (Linear Regression)," the author, who was tracking the price of Ethereum cryptocurrency, recorded daily transactions vs price three times per month for 3 months, from 1/1/2021 through 3/20/2021, and put these observations in a table. Unrelated to the table, the author also noted that, "Now, one year ago on 1st March 2020, the number of daily transactions was 628,500. We have now doubled, to around 1.2M daily transactions in March 2021. It is likely that we will at least double that again to 2.4M daily transactions by March 2022."

Using the transactions vs price data and the estimated transactions amount for March 2022, the author used regression to predict that the price of Ethereum in March 2022 would be $6,414. In fact, in March 2022, the price of Ethereum was about $3,000, less than half the amount predicted (As of April 9, 2023, the price was $1,840.) What went wrong?

Although there was a positive correlation between transactions and price, the scatterplot and regression line were from 2021 data, whereas the prediction was for March 2022, far outside the initial data range. Things don't stay linear forever, especially in financial markets. They often level off or drop, ruining the initial linearity. Also, the transactions value used for the prediction was based on a sketchy estimate in which the author was trying (unsuccessfully, as it turned out) to be a good prognosticator. This further illustrates limitations of regression analysis.

Some Things aren't Linear

Not all data can be modeled with a straight line. For scatter plots that don't follow a straight line, methods other than the correlation coefficient and regression line can be used.

pixssa/Shutterstock.com

Fuel Efficiency

A gas powered car's fuel efficiency in miles per gallon (mpg) changes according to the car's speed (it's different for electric cars). When the car first starts, its fuel use is high, so mpg is low. As the car's speed increases, mpg increases as well until the car reaches optimal speed, near 55 mph in most cars. As speed continues to increase, fuel use picks up again and mpg decreases. Here are hypothetical data and scatter plot with a polynomial regression line.

Inked Pixels/Shutterstock.com

speed	mpg
10	20
20	25
25	28
35	32
47	35
60	28
70	23

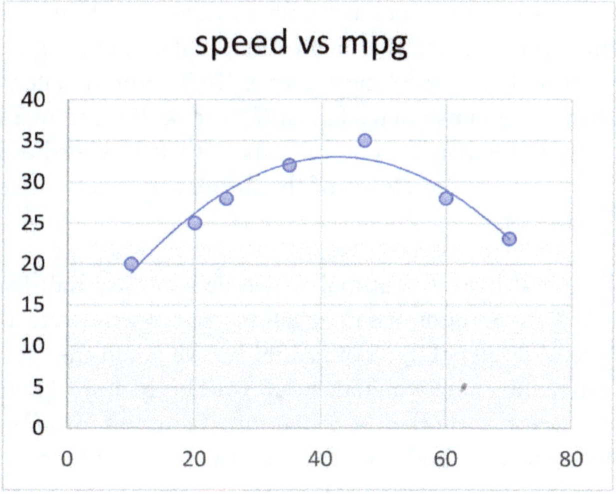

The relationship between speed and mpg is not linear but "polynomial." Polynomial functions are a general class of functions of which linear functions are the simplest. Data that can be modeled by non-linear polynomial functions need to be analyzed using methods other than the correlation and regression methods discussed here, but the basic ideas are the same.

Two Half-Lives Do Not Make a Whole Life

The half-life of a drug is the time it takes for the amount of the drug's active substance in your body to diminish by half. For example, the blood pressure

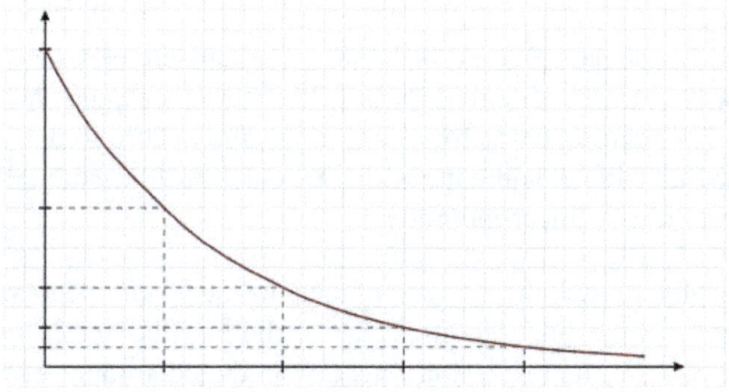

MilanB/Shutterstock.com

medication atenolol has a half-life of about 6 hours. This means that 6 hours after you've taken the medication, half of the drug's active substance has left your body or is no longer active. Then, 6 hours later, half of what's left of the drug is no longer active, meaning that at the end of 2 half-lives, ½ * ½ = ¼ or 25% of the drug's active substance is left in your body. After 3, half-lives, ½ * ½ * ½ = 1/8 or 12.5% of the drug's active substance remains in your body. And so on.

Two half-lives do not make a whole life. Two half-lives would make a whole life if the amount of a drug left in your system were a linear function of time. Then, for atenolol, 6 hours after you've taken the medication, ½ of the drug's active substance would have left your body, and 12 hours after you've taken the medication, another ½ of the initial amount, or *all* of the medication would have left your body, which is not the case. Mathematically, the difference is ½ * ½ (non-linear), vs ½ + ½ (linear).

Here is the graph of half-life of atenolol vs hours from the time the drug is taken.

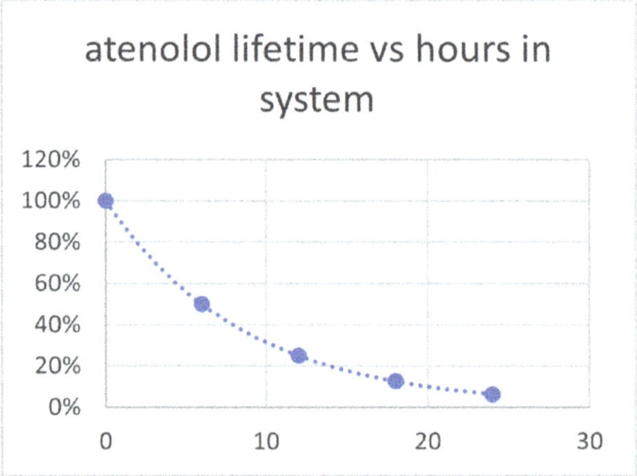

The curve in the scatter plot is not linear. The half-life of a pharmaceutical product is a type of "exponential decay." The curve is *exponential.* Half-life tends to vary depending on factors such as age and health, so data analysis to determine the half-life of someone with particular health issues taking atenolol would rely on non-linear methods.

With the half-life model, the amount of drug in the system decreases, but the rate of decrease levels off. The first part of the half-life graph looks almost linear but if you use the regression line, the predicted amount of the drug in the user's system soon becomes negative, which doesn't make sense. Here's the atenolol scatter plot with the regression line added. Notice that when x > 22.3 hours, the regression line predicts a negative percent of atenolol in your body.

One of the biggest misuses of linear regression is using the regression line for data that is not linear. For situations such as this, even if data is linear for part of the scatter plot, it won't necessarily stay that way. The "least squares" method used for obtaining the linear regression line can be used for any type of line provided you have the formula for the line and a scatter plot. The linear version is the simplest, both computationally and conceptually.

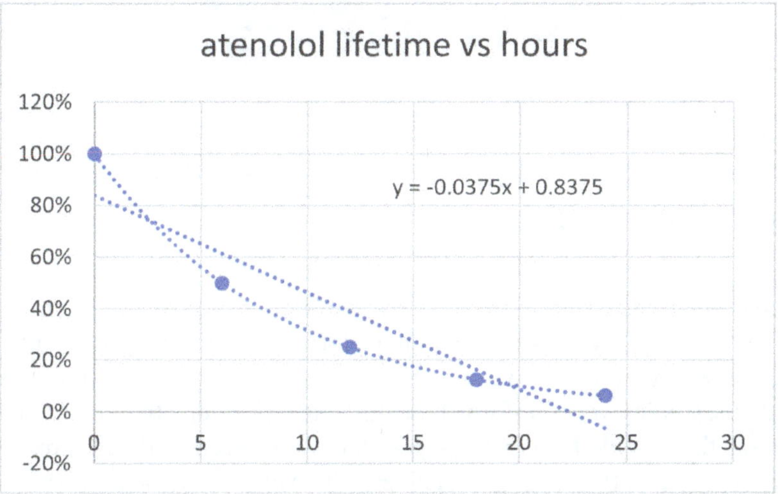

Another type of exponential decay is radioactive decay. Radioactive decay measures the time of disintegration of a radioactive substance. Just like the half-life of a pharmaceutical product, radioactive decay is measured in half-life. For example, the half-life of radioactive iodine-131 is 8 days. Here is a graph of the half-life of radioactive iodine. Again, this is not linear, so analyzing the decay of a radioactive substance should be done using methods other than linear regression.

Satakorn/Shutterstock.com

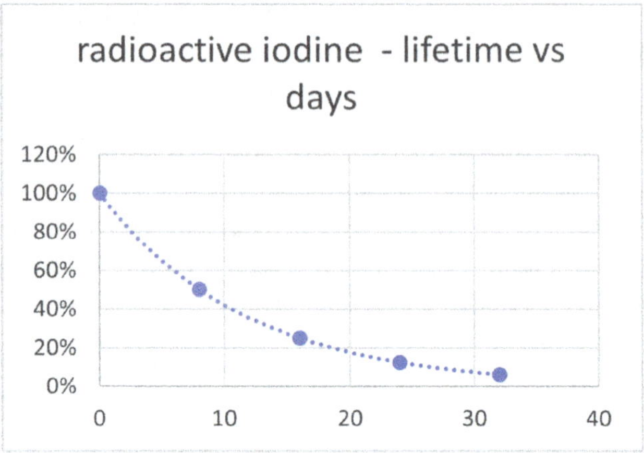

Good Things (Like Regression) Don't Last Forever

The regression line is based on a particular scatter plot. If you try to extend the line past the x-values of the scatter plot, you're looking for trouble. For example, when the U.S. and Russia first started sending satellites into space in the 1950s, studies were done involving satellite technology that tried to predict what satellites would be like in the year 2000. One such study used regression analysis based on the weight of satellites, which appeared linear for the data in the study. Unfortunately, things like satellite weights level off over time. The study didn't take the eventual leveling off into account and predicted that the weight of a satellite in the year 2000 would be greater than the weight of Earth.

Song of the Sirens

Regression depends on the scatter plot. Things change. The regression line was never intended to predict Y when X is beyond the range of the x-values in the scatter plot. Sometimes it takes a bit of discipline to adhere to this simple rule because the regression line is easy to use and will predict Y for *any* value of X, whether things stay linear or not. Simplicity is not the same as accuracy. The regression line, useful in many situations, can be compared to the Sirens of Greek mythology: Dangerous creatures who lured statisticians with their

Catmando/Shutterstock.com

enchanting music and beautiful singing to shipwreck on the rocky, non-linear (fractal) coast of their island.

Bottom Line

The regression line, or least squares line, is an optimal way to predict one variable from another in a linear way. It is the best fitting straight line to a scatterplot and is one of the most popular prediction methods. The regression line should only be used with correlated data and only within the range of data in the scatterplot on which the line is based. Linear relationships are typically only linear in a certain range and do things like "level off" after a while.

CHAPTER 23

CONFLICT OF INTEREST RATES

"It was the best of times. It was the worst of times."
Charles Dickens, A Tale of Two Cities

According to a June 15, 2024, Washington Post article titled, "Fed's higher interest rates do not get all the credit for lower inflation," David J. Lynch says, "Improvements in supply chains and availability of labor have helped, even though the economy didn't respond as expected to rate hikes." The "expected" response to rate hikes was increased unemployment. This was spelled out by a well-known economist, who said in a speech at the London School of Economics, "We need five years of unemployment above 5% to contain inflation -- in other words, we need two years of 7.5% unemployment or five years of 6% unemployment or one year of 10% unemployment." The Fed Chair followed up with a statement from Jackson Hole, Wyoming on August 26, 2022, according to a story in the Wall Street Journal, "While rate increases would bring down inflation, they will also bring some pain to households and businesses."

In fact, interest rate increases didn't increase unemployment, and there's no proof that they decreased inflation (there were lots of other likely factors). But they did cause pain.

When Congress established the Federal Reserve in 1913, they set three key objectives for the Fed's monetary policy: "maximizing employment, stabilizing prices, and moderating long-term interest rates." The Fed Chair has stated numerous times that the objective of raising interest rates was to increase unemployment. How is that "maximizing employment," as stated in the Fed's guidelines?

The Fed Chair thought that raising interest rates was the best way to bring down inflation, largely because of the Phillips Curve, which had been discredited by the time he used it as a statistical underpinning of Fed policy. The Phillips Curve was originated by William Phillips, a New Zealand economist who wrote a paper in 1958 called "The Relation between Unemployment and the Rate of Change of Money Wage Rates in the United Kingdom, 1861–1957." The curve shows an inverse relationship (negative correlation) between unemployment and inflation in the United Kingdom from 1861 to 1913. Phillips found that the inverse relationship remained when the data was extended to 1957.

The Phillips Curve shows that as unemployment goes up, inflation (rate of change of prices) comes down, on average. This has been interpreted by the Federal Reserve to mean that if you want to bring inflation down, you should increase unemployment (violation of Fed policy?), which you can presumably do by raising interest rates. This is convenient, because raising (or lowering) interest rates is one of the few things the Fed can do. Like the old adage, "If your only tool is a hammer, treat every problem like a nail." If a huge asteroid was hurtling towards Earth, the Fed Chair would probably say that raising interest rates was the only way to stop it.

In addition to using 19th century data from the United Kingdom to model 21st century inflation in the U.S., there's a problem with using Phillips curve as a "causative" model. The data is observational (not a controlled experiment), and, although there may be a correlation between unemployment and inflation, observational data shouldn't be used to infer causation.

As we discussed in the Correlation chapter, in the late 1950s, when the polio vaccine was being developed, someone noticed that there was a positive correlation between polio incidence and soft drink sales: A researcher concluded that consuming soft drinks *causes* polio. As we saw, there was a background variable that was responsible for the correlation: The weather. In warm weather, more soft drinks are consumed, and the polio virus is more active. Just like the foolish researcher who thought soft drink consumption caused polio, the Fed apparently thought that raising interest rates would cause higher unemployment, which would then cause a decrease in inflation due to the Phillips curve. It didn't happen.

When interest rates are raised, there's lots of pain. Mortgage rates go up. Young people and working people who can buy a house when interest rates are moderate or low can no longer afford to do so. The same is true for automobile loans and credit card debt. When interest rates are low, it's easy for small businesses to borrow money, which is necessary for growth and even survival for some businesses, like restaurants, where the effects of Covid are still being felt. What's the point of all this pain for a policy that didn't make statistical sense to begin with?

According to a 2020 article by James Dorn, published by the Cato Institute:

> "The history of economic thought is replete with examples of economic ideas that are politically attractive and persist even after they have little empirical support or theoretical validity. That has certainly been the case with the Phillips curve."

Dorn goes on to say:

> "The underlying problem is that the Phillips curve misconstrues a supposed correlation between unemployment and inflation as a causal relation. In fact, it is changes in aggregate demand that cause changes in both unemployment and inflation. The Phillips curve continues to misinform policymakers and lead them astray."

Aggregate demand was the background variable here, just like the weather was behind the relationship between soft drink consumption and polio. In spite of this, the Fed is still using the Phillips Curve. They could just as well be making economic forecasts by interpreting sheep entrails, similar to what was done by ancient oracles, the predecessors to today's economists.

The one area where increased interest rates may have had an effect in reducing inflation is the housing market. But even that's not clear. In a June 6, 2024, Washington Post op-ed piece, "Note to Fed: It's okay to cut interest rates now," economists Jim Parrott and Mark Zandi point out a flaw in the definition of inflation with regard to how housing is factored into the inflation formula. When this is corrected, the need to keep interest rates high disappears. This points out another problem. The definition of "inflation" is arbitrary to some degree and may not accurately represent current economic conditions.

It's natural to ask why the Fed would still rely on a discredited statistical model based on data from 19th century UK, especially since increased interest rates have not done what they were supposed to do. It's seldom true that a policy hurts everyone. Maybe there are some who benefit. For example, there are those who have money to put in savings accounts and who benefit from higher rates. In a Forbes article, Jason Kirsch notes,

> "As interest rates rise, the interest income from loans typically increases faster than the interest paid on deposits, leading to wider profit margins. Additionally, higher interest rates can boost the earnings of insurance companies and investment firms, as they often hold large portfolios of interest-sensitive assets."

Insurance companies and investment firms profit when interest rates go up. The Federal Reserve is self-funded and largely supported by the interest from the securities it owns. So if interest rates are high, the Federal Reserve can make bigger profits for their own funding. This seems to be a conflict of interest rates. I'm reminded of the famous quote by Upton Sinclair: "It is difficult to get a man to understand something when his salary depends on his not understanding it."

Bottom Line

Economists and others who are in charge of keeping the economy on track should follow the rules of basic statistics and know that correlation does not mean causation.

CHAPTER 24

THE NORMAL CURVE

What's so special about the bell shaped, or normal curve, the probability distribution that seems to show up everywhere? First: A probability bar graph is a bar graph with bars of equal widths, and relative areas representing probabilities for the bars. A probability bar graph can sometimes be approximated by a smooth curve, like the bell shaped curve shown above, called a probability curve, with the curve and bar graph scaled so that the total area under the curve equals 1. Many, but not all, probability bar graphs, along with graphs of data obtained from random samples, look roughly bell-shaped and can be outlined with a particular curve called the "normal" curve.

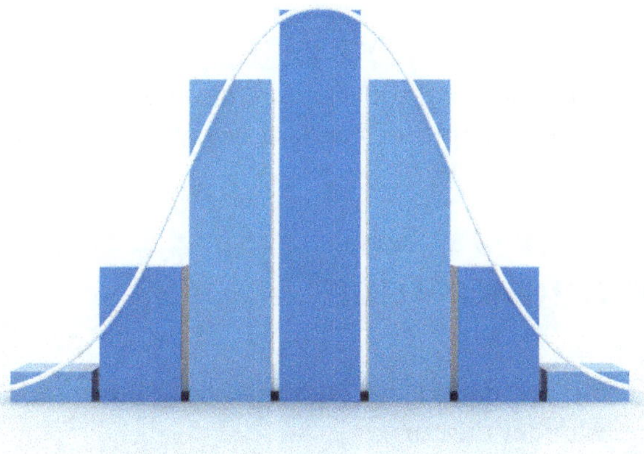

Ersin Kurtdal/Shutterstock.com

Data that arises from some random samples have summary statistics whose probabilities can be approximated by the normal curve. This is because of a famous mathematical result called the Central Limit Theorem, which says that for a large enough random sample, probabilities for averages and proportions can be approximated by the normal curve. This means that the normal curve is sometimes a good fit to the probability graphs of the appropriate distributions. For example:

- Counting successes in a large number of independent trials, where each trial ends in either success or failure and where the success probability is the same from trial to trial.
- Heights, weights, and other physical measurements.
- Sums, averages, and proportions from a random sample with a large sample size
- Errors made by a scale or other measurement device

The normal curve has a specific mathematical formula. It is symmetric about its mean. The total area under the curve equals 1, the same as any probability curve. There is a normal curve for every possible mean and standard deviation. Once the mean and standard deviation are determined, so is the curve. (Standard deviation is a common measure of variation used in statistics. Roughly, the standard deviation is the average distance values are from the mean, or average.)

> A normal curve can be rescaled in units of standard deviation from the mean.
>
> The result is a generic normal curve called the "standard" normal curve, or Z.
>
> The Z curve has mean = 0 and standard deviation = 1. Z represents all normal curves in units of standard deviation from the mean.

The Empirical Rule

The "empirical rule" for the normal curve provides three useful probabilities, expressed as areas under the curve:

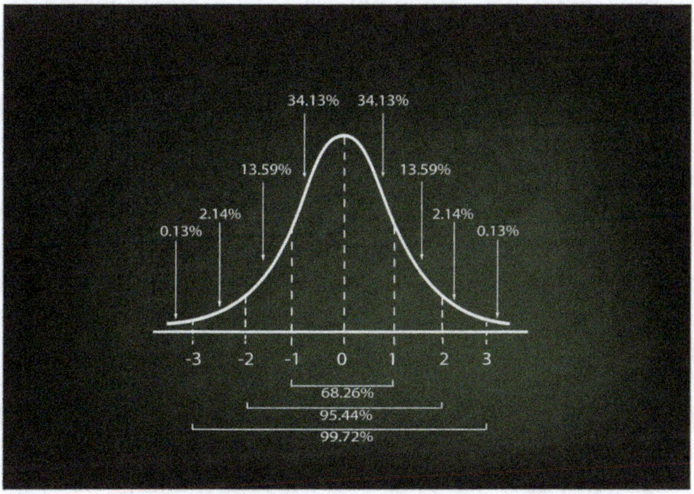

Iamnee/Shutterstock.com

1. About 68% of normal values are within one standard deviation of the mean. Since Z is scaled in units of standard deviation from the mean, there's a 68% chance that a randomly selected Z value is between -1 and 1.
2. About 95% of values are within two standard deviations of mean. Equivalently, there's a 95% chance that a randomly selected Z value is between -2 and 2.
3. Almost all values (99.73%) are within three standard deviations of mean. There's a 99.7% chance that a randomly selected Z value is between -3 and 3. As we shall see, this is deceptive, since it causes the mistaken belief that you will never see Z scores greater than 3 or less than -3.

Smooth Curves and Not-So-Precise Measurements

When you're using a smooth, or continuous curve to approximate a probability histogram, probabilities are expressed as areas under the curve for whatever interval of values you need to know. With a smooth curve, the probability of a particular value equals 0, because there is no area for the straight line above a point. There is, however, area under the curve for a small interval of values around the point.

To find the probability that a randomly selected Z value is between 0.9 and 1.1, you find the area under the standard normal curve for the range of Z values between 0.9 and 1.1. To find the probability that a randomly selected Z value is between .95 and 1.05, you find the area under the Z curve for Z values between .95 and 1.05. As we close in on Z = 1 we find areas under the curve for smaller and smaller intervals containing 1. Since Z is a smooth curve, the areas of these smaller and smaller intervals containing 1 get smaller and smaller and so do the corresponding probabilities. As the intervals grow smaller, the area under the curve gets closer and closer to zero until the intervals collapse to the point Z =1, and the probability that a randomly selected Z value equals exactly 1 becomes zero.

This makes perfect sense. When we're dealing with continuous data, like physical measurements, we are limited by the precision of our measuring devices. If we say that someone is 6 feet tall, then if we are rounding off to the nearest inch, we mean that the person is between 5'11.5" and 6'0.5". If we want to be more precise and if we have a good enough measuring device, we could round off to the nearest ½ inch, or the nearest quarter inch, etc. But no matter how good our measuring device is, we'll always be rounding off to *something*. Who cares if someone is really 5.99999 feet tall or 6.00001 feet tall? We can just use 6 feet tall and call it a day. Or if you care, you can look for a measuring device that will handle the level of precision you desire. Warning: There's no measuring device that can measure something to infinite precision. Even a "nano-ruler" would eventually run into the limitations imposed by quantum mechanics.

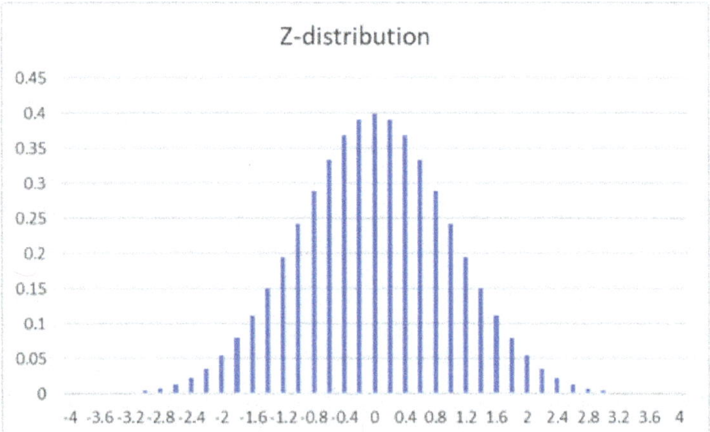

It's important to note that even though typical Z graphs make it look like Z essentially hits the horizontal axis at around 3 and -3, the graph never actually hits the axis. In other words, no matter how far out you go in either the positive or negative direction, there is always tiny bit of area, that is, a tiny bit of probability left. This is true for all normal curves. The above example gives a glimpse of how small this area is. To find the positive part when computing $P(Z \geq 7)$, you have to go out to the 12th decimal place, so for most practical purposes $P(Z \geq 7)$ and $P(Z \leq -7)$ equal 0. If you see a value like this when looking at data, it's an outlier, that is, weird. However, the corresponding probability is technically not zero, nor is any other region under the curve, although the tail probabilities are so small, they're typically rounded off to zero.

OTHER NORMAL CURVES

Since the Z curve has mean = 0, and standard deviation = 1, it can be used to represent any normal distribution, X, expressed in units of standard deviation from the mean. The conversion to units of standard deviation from the mean is done using the formula

$$Z = \frac{X - \text{mean}}{s.d.}$$

Here's an example using weights of turkeys, which, since these are physical measurements, can often be approximated using the normal curve.

THE TURKEYS

On a certain farm, weights of adult turkeys are normally distributed, with mean = 12 pounds, standard deviation = 2 pounds.

Let X = weight of randomly selected turkey. Then, using the Excel function "norm.dist," we find: $P(X \geq 16 \text{ pounds}) = 0.0228$

veleknez/Shutterstock.com

There's about a 2% chance that a randomly selected adult turkey from this farm weighs more than 16 pounds. Using the empirical rule for this population of turkeys, we find that

> About 68% of turkeys weigh between 10 and 14 pounds (within 1 s.d. of the mean)
>
> About 95% of turkeys weigh between 8 and 16 pounds (within 2 s.d.'s of the mean)
>
> Almost all turkeys weigh between 6 and 18 pounds (within 3 s.d.'s of the mean)

Still looking at weights of turkeys, we find that for a randomly selected turkey, X,

> $P(X \leq 6 \text{ pounds}) = 0.0014$
>
> $P(X \geq 22 \text{ pounds}) = 0.0000003$ or about 1 in 3.5 million.

A turkey that weighs 22 or more pounds is an outlier. It's very unlikely that a turkey from this farm weighs that much. If you think you have found such a bird, maybe it's an emu who sneaked into the flock when nobody was looking. Or maybe the normal curve isn't an accurate probability model for the turkeys, something people find hard to accept.

Bottom Line

Due to the central limit theorem and other mathematical results, the normal curve is one of the most important and oft-used probability distributions in statistics. Because of its ubiquity and ease of use, the normal curve is also misused, usually when the conditions for using it aren't fully met. The normal curve was developed a few centuries ago when there were no computers and when mathematicians needed methods like this in order to complete their analyses. In the future, software, both analytical and simulations, will make the normal curve less important as part of the statistician's toolkit, but for now, it's a mainstay of probability and statistics.

CHAPTER 25

THE BINOMIAL AND GEOMETRIC DISTRIBUTIONS

THE MULTIPLICATION RULE

Consider a combination lock that is common on bike locks and shown in the picture above. The lock has 10 choices, 0 through 9, for each of 5 reels. The user sets one configuration of numbers that opens the lock. How many configurations are there altogether? The multiplication rule states that since there are 10 choices for each reel, 0 through 9, and since there are 5 reels, there are $10*10*10*10*10 = 10^5 = 100{,}000$ configurations.

SSS55/Shutterstock.com

The combination lock is a simple but ingenious device that lets you select one configuration out of 100,000 by setting the reels in a particular order. You can use the multiplication rule for independent events in probability, as well. Suppose a thief tries to figure out the correct configuration of numbers to open your lock and picks a number at random on each reel. The chance of getting the correct configuration equals $(1/10)^5 = 1/100{,}000$. If the thief tries one random combination per minute 24 hours per day, 7 days per week, it will take an average of more than two months until the code is broken, and your bike is stolen. Of course, the thief could use a cable cutter and not worry about the lock (This happened to me).

FACES

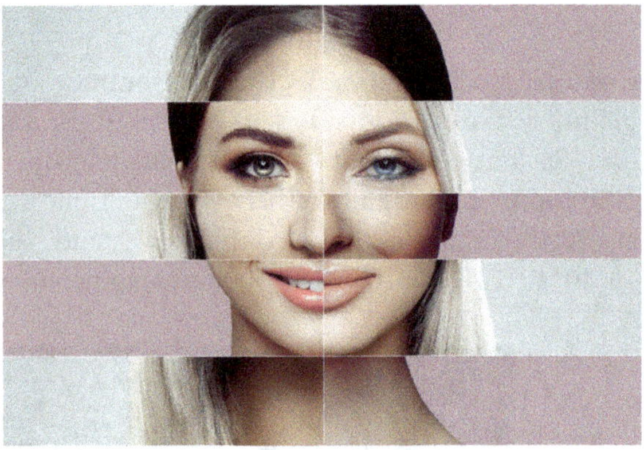

Artmim/Shutterstock.com

In the above picture, a face is segmented into 10 segments, like a puzzle. There are right and left parts for hair, eyes, nose, mouth and neck. There are 6 facial choices for each segment. By the multiplication rule, there are $6^{10} = 60{,}466{,}176$ faces that can be made from these 10 segments. Breaking a face into segments and looking at configurations is the basis for facial recognition software. You don't have to look at 60 million different faces. Instead, you find different faces by making choices for each segment. Face segmenting software is also used by plastic surgeons to help select an appropriate look for a patient.

GETTING YOUR DUCKS IN A ROW

Did you ever try to get your ducks in a row? Have you ever wondered how many ways there are to do this? Suppose you have 15 ducks. How many dif-

kumikomii/Shutterstock.com

ferent ways can you arrange them in a row, assuming that you can tell them apart, maybe with name tags. Such an arrangement, or ordering of ducks (and other things), is called a *permutation*. You can permute ducks in a row, books on a shelf, concertgoers in a line, sequences of letters, numbers, whatever. Two is the minimal number of things you can permute. Let's start with two letters, A, B, or two ducks, with name tags A and B. (You can't permute one thing, like the letter A, because it's just standing there, by itself. To quote the song by Harry Nilsson, "One is the Loneliest Number You Will Ever See.")

It's easy to see that there are two permutations of the letters A and B: AB, BA. What about three letters A, B, C? Any one of the three letters can start an ordering. Once the first letter is selected, either of the two remaining letters can follow it. Once the first two letters are selected, there is only one letter left. There are 3 * 2 * 1 = 6 permutations of A, B, C:

ABC, ACB, BAC, BCA, CAB, CBA.

What about four letters: A, B, C, D? These are just symbols. You can make them into name tags for 4 ducks and ask the same question about ducks or anything else for that matter. The same reasoning applies. There are four starting letters, three letters that can follow each starting letter, two letters available for each starting sequence of two letters, and one letter left to end a sequence. This gives a total of 4 * 3 * 2 * 1 = 24 permutations of the letters A, B, C, D:

ABCD, ABDC, ACBD, ACDB, ADBC, ADCB, BACD, BADC, BCAD, BCDA, BDAC, BDCA, CABD, CADB, CBAD, CBDA, CDAB, CDBA, DABC, DACB, DBAC, DBCA, DCAB, DCBA

A pattern emerges. For the five letters A, B, C, D, E, there are 5 * 4 * 3 * 2 * 1 = 120 permutations. There are 6 * 5 * 4 * 3 * 2 * 1= 720 permutations of six things, and so on. It quickly becomes impractical to write down all the permutations or even to write down the permutation formula. There is, however, a convenient way to write the permutation formula. A whole number followed by an exclamation point is called "factorial" and denotes that number multiplied by every number less than it down to 1. For example, 6! read "6 factorial," equals

6 * 5 * 4 * 3 * 2 * 1 = 720.

Likewise, 7! = 7 * 6 * 5 * 4 * 3 * 2 * 1= 5,040.

There are 8! = 40,320 permutations of the words in the sentence, "Is it not yet time for my painkiller?"

There are 9! = 362,880 permutations of nine people in a line.

There are 10! = 3,628,800 permutations of ten plants on a shelf.

There are 11! = 39,916,800 ways to play eleven different Lady Gaga songs.

There are 15! = 1,307,674,368,000 permutations of 15 ducks in a row. That's more than 1.3 trillion permutations, a tall ordering, so to speak.

Stephen Chung/Shutterstock.com

COMBINATIONS

There are 15 ducks in a row. Four of the ducks step away to have a conversation. How many combinations of ducks can you select when you choose 4 ducks from 15? Now instead of picking one number from each of several reels on a combination lock, you can think of all the balls as being in one box and drawing a few out. Or all the ducks being in a line and selecting a few to quack at each other (Give them some cheese and you'll have cheese and quackers).

When you select a few ducks from a line of ducks, there is still multiplying to do, similar to the permutations formula. As we saw, if there are 15 ducks in a row, there are 15! or about 1.3 trillion ways to arrange them in a line. How many ways are there to select 4 ducks from the 15? We can use the same type of reasoning as with permutations: There are 15 ways to select the first duck. After we've selected the first duck there are 14 ways to select the second duck. After we've selected the second duck there are 13 ways to select the third duck. And after we've selected 3 ducks, there are 12 ducks to select the 4th duck. We can put the selected ducks in a row, with the first duck selected going first, the second duck going second, the third duck going third and the fourth duck going fourth. There are 15*14*13*12 = 32,760 ways to do this.

There's a problem. The ducks were selected to have a conversation, not to put in a smaller line. The order of selection isn't important. We can remove the order of selection by noting that for each ordered selection of 4 ducks, there are 4! permutations, or ways to arrange them. So, if we divide the number of ordered ways to select 4 ducks by 4! we will have the number of unordered ways to choose 4 ducks from 15 to have a conversation:

$$\frac{15*14*13*12}{4*3*2*1}=1{,}365$$

There are 1,365 ways to choose 4 ducks from 15 without worrying about the order.

This formula is used to find the number of ways you can "choose" or select objects from a larger group. The formula is cumbersome if you write down all the factors, but we can use the factorial formula to simplify it. For the ducks, we can rewrite 15*14*13*12 / 4*3*2*1 as 15! / 4!*11! The 11! in the denominator cancels out the 11! that's part of the 15! to give the desired result. This formula comes up so much that there's a special notation: 15! / 11!*4! =

C(15,4), read "the number of combinations of 15 things taken 4 at a time," or, more succinctly, "15 Choose 4", the number of ways to choose 4 things from 15. The general formula looks like this:

C(n, k) = n! / k!*(n-k)! = the number of ways to choose k things from n. This is called the *combinations formula.*

We already saw that there were 1,365 combinations when choosing 4 ducks from 15. It follows that there are 15! / 7!8! = 6,435 ways to choose 7 ducks from 15, as well as to choose 8 ducks from 15. If you choose 7, you've also chosen 8, namely the ducks who were not chosen.

The Excel function for the combinations formula is "combin(number, number chosen)". Using this function, the number of ways to choose 4 ducks from 15 equals combin(15,4) = 1,365, and the number of ways to choose 7 ducks from 15 equals combin(15,7) = 6,435.

There are C(20,5) = 20! /5!15! = 15,504 ways to choose 5 ducks from 20. So, if you randomly select 5 ducks from 20, the chance equals 1/15,504 that you will select the 5 oldest ducks.

The Mega Millions Lottery Revisited

Recall that a Mega Millions lottery ticket consists of 5 numbers from 1through 70 and a Mega Number from 1 through 25. At the lottery drawing 5 winning numbers from the numbers 1 through 70 are randomly selected, along with a random winning Mega Number from the numbers 1 through 25. The number of ways to choose 5 numbers from 1 through 70 equals C(70,5) = combin(70,5) = 12,103,014. We multiply this by 25, the number of ways to choose 1 Mega Number from the numbers 1 through 25, to get

SNew Africa/Shutterstock.com

12,103,014 * 25 = 302,575,350. You may recognize as the number of possible Mega Millions ticket combinations. This makes a single ticket holder's chance of winning the Mega Millions jackpot equal to 1/302,575,350. Now you know where that number came from.

The King's Lottery

We also discussed the King's Lottery, in which you pick 20 numbers from 1through 80. At the lottery drawing, 20 winning numbers are randomly selected from the numbers 1through 80. You win if all 20 of your numbers match the 20 winning numbers. There are C(80,20) = combin(80,20) = 3,535,316,142,212,180,000 = 3.54 quintillion ways to pick 20 numbers from 80 (It's not a typo). Your chance of winning equals 1/C(80,20) = 1/3.54 quintillion. As discussed earlier, that's why nobody won.

Poker Hands

The combinations formula can be used to find the number of ways to get various poker hands. Since a poker hand is defined by five cards, let's start

Andrew Angelov/Shutterstock.com

by asking how many possible poker hands there are. Since a deck of cards has 52 cards in it, the number of poker hand combinations equals C(52,5) = 2,598,960. Let's try something a little more complicated. How many ways are there to get a flush, namely all cards of the same suit? Consider a spade flush. Since there are 13 spades in a deck, there are C(13,5) = 1,287 ways to choose a spade flush. Since there are 4 suits, hearts, clubs, diamonds, and spades, there are 4 * C(13,5) = 5,148 flushes. This includes straight flushes, which are a different poker hand, so we must subtract them from the total. A straight flush is all five cards in sequence and all of the same suit, like 3,4,5,6,7, all hearts, or 10, Jack, Queen, King, Ace, all spades (called a royal flush). There are 10 possible straight flush sequences (starting card can be Ace through ten) and 4 suits so there are 10 * 4 = 40 straight flushes, leaving a total of 5,148 – 40 = 5,108 flushes.

A full house is when you get three cards of one denomination and two of another. Since there are C(4,3) ways to pick three cards out of four and since there are13 denominations, there are 13*C(4,3) = 52 ways to have the 3-card component of the full house. Once the 3-card denomination is chosen there are 12 denominations left for the 2-card pair. For each denomination there are C(4,2) = 6 two card pairs, so there are a total of 12*6 = 72 two card possibilities. This gives 52*72 = 3,744 full house combinations. Here, we have used both the combinations formula and the multiplication formula.

To find probabilities for being dealt a specific poker hand when 5 cards are dealt at random from the deck, just take the number of hands of a specific type and divide by C(52,5) = 2,598,960. For the hands we've just analyzed,

P(straight flush) = 40/2,598,960 =0.0000154

P(full house) = 13*C(4,3)*12*C(4,2)/C(52,5) = 3,744/2,598,960 = 0.00144

P(flush) = (4*C(13,5)-40)/C(52,5) = 5,108/2,598,960 = 0.00197

There are some poker games, like Texas Hold-em, that involve putting together your best poker hand out of more than five cards, so even though the combinations we've just discussed remain the same, the probabilities of ending up with certain hands become more complicated.

Again With the Coin Tossing

OB production/Shutterstock.com

We've been talking about coin tossing and related experiments from time to time to time. Everything was fine until the number of tosses grew large. The combinations formula handles this. For example, suppose you toss a coin 10 times. What's the probability that you get 5 heads and 5 tails in some order? 6 heads and 4 tails? 7 heads and 3 tails? As we saw earlier, in ten tosses, there are 2^{10} = 1,024 different possible configurations. This follows from the multiplication rule and serves as the denominator for resulting probabilities. To find the number of coin-toss results with 5 heads and 5 tails, we just use the combinations formula to get: C(10,5) = combin(10,5) = 252. So,

P(5 heads in 10 tosses) = 252/1024 = 0.246

P(6 heads in 10 tosses) = C(10,6)/1,024 = 210/1,024 = 0.205

P(7 heads in 10 tosses) = C(10,7)/1,024 = 120/1024 = 0.117

Even with the combinations formula at our disposal, however, computational problems arise. For example, suppose we toss a coin 30 times and want to know the probability of getting 20 or more heads. Since 20 or more heads means 20 heads or 21 heads or 22 heads ...or 30 heads, we need to compute the probability of getting 20 heads plus the probability of 21 heads ... plus the probability of 30 heads. Even if we use the Excel function, *combin*, it will take a few minutes to put things together. However, if we generalize the situation, there is another spreadsheet function that will handle this computation effortlessly.

THE BINOMIAL DISTRIBUTION

igor kisselev/Shutterstock.com

The combinations formula allows us to compute probabilities for many situations, one of which is the binomial probability distribution. The binomial distribution consists of a sequence of independent trials where on each trial only two things can happen, "success" or "failure," where the success and failure probabilities are the same from trail to trial, and where the binomial variable gives the number of successes in n trials. These labels are arbitrary and refer only to the fact that there are just two things that can happen (hence the term "*bi*nomial"). Coin tossing has a binomial distribution because tosses are independent and the two things that can happen on each trial are heads and tails, with the probabilities of heads and tails staying the same from trial to trial (We don't consider the possibility that the coin might land on its side, although apparently such things have happened).

We already know the formula for coin tossing probabilities. One of the things that makes coin tossing easier to analyze than other binomial probabilities is that there's a 50-50 chance that heads, or tails, comes up on each toss. With the general binomial distribution, even though there are two possible results on each trial, the results of a trial don't have to be 50-50, as the following example shows.

INTERNET ADVERTISING

If someone clicks on a particular Internet ad, there is a 20% chance that they will make a purchase. In 10 clicks, what is P(3 purchases)? This is different from tossing a coin 10 times because the chance of a purchase isn't 50%, like the chance that heads comes up. Instead, it's 20%. The chance of 3 purchases

Rawpixel.com/Shutterstock.com

isn't C(10,3) / 2^{10} because different configurations have different probabilities depending on the number of purchases. However, by the multiplication rule, every specific sequence of 10 clicks resulting in three purchases and seven non-purchases has the same probability regardless of the order of the purchases, namely, $0.2^3 * 0.8^7$, where 0.2 is the chance of a purchase on a given click, and 0.8 is the chance of no purchase. All you have to do is multiply this probability by the number of combinations of 10 clicks that have 3 purchases and 7 non-purchases. That would be C(10,3) = 10!/3!*7! = 120. So, since each sequence of 3 purchases and 7 non-purchases has probability = $0.2^3 * 0.8^7$, we add the number of such sequences by multiplying by C(10,3) to get

P(3 purchases in 10 clicks) = C(10,3)*$0.2^3 * 0.8^7$ = 0.201. Likewise,

P(4 purchases in 10 clicks) = C(10,4) *$0.2^4 * 0.8^6$ = 0.088

And so on.

The General Binomial Formula

We're ready for the general binomial formula.

There are n independent trials. On each trial, there are two possibilities: success or failure, where P(success) = p, and P(failure) = q =1-p, and where the success probability remains the same from trial to trial. Let X = number of successes in the n trials. Then

$$P(X = k) = C(n, k) * p^{k} * q^{n-k}$$

Also, $P(X \leq k) = \Sigma P(X=i)$ for $i = 0, 1, \ldots k$, where Σ means sum.

The spreadsheet function, "binom.dist" computes binomial probabilities.

Back to Internet advertising. Recall that the probability equals 0.2 that someone will make a purchase if they click on a particular Internet ad. In 30 clicks, find the probability of 3 or fewer purchases. Using the general binomial formula with p = 0.2, q = 0.8, and n = 30, and letting X = number of purchases, find $P(X \leq 3)$. Computations get tedious, so we can use the Excel function, "binom.dist", which, as usual, returns binomial probabilities of two types, cumulative and exact, which are indicated in binom.dist by "true" (cumulative), and "false" (exact). The function is of the form

=binom.dist(successes, trials, success probability, true).

In the internet advertising example with n = 30, p = 0.2 and q = 0.8, we find

$P(X \leq 3)$ =binom.dist(3, 30, 0.2, true) = 0.1227

To find the probability of exactly 3 purchases in 30 clicks, P(X = 3), rather than $P(X \leq 3)$, enter =binom.dist(3, 30, .2, false) = 0.0785

Land of the Flea

In a certain area, 40% of dogs have fleas. 20 dogs are selected at random. Let X = number of dogs in the sample with fleas, then

$P(X = 10) = C(20,10)*0.4^{10}*0.6^{10}$ = binom.dist(10, 20, 0.4, false) = 0.1171

$P(X \leq 9)$ =binom.dist(9, 20, 0.4, true) = 0.7553

$P(X \geq 13) = 1 - P(X \leq 12)$ = 1 – binom.dist(12, 20, 0.4, true) = 0.02

Daniel Myjones/Shutterstock.com

Back to Coin Tossing

Since coin tossing is a special case of the binomial distribution, we can rewrite the coin tossing formula to match the formula we're now using. Using the combinations formula, we saw that P(5 heads in 10 tosses) = C(10,5)/1,024 = 252/1024 = 0.246.

In one toss, P(H) = P(T) = 0.5, so in ten tosses any specific sequence of 5 heads and 5 tails has probability equal to $.5^{5}*.5^{5} = .5^{10}$ =1/1,024. So, we can rewrite the coin tossing formula as P(5 heads in 10 tosses) = C(10,5)/1,024 = C(10,5) $*.5^{5}*.5^{5}$, which matches the binomial formula. For example, using this formula, we find that the probability of getting 6 heads in 10 tosses equals C(10,6)/1,024 = C(10,6) $*.5^{6}*.5^{4}$ = binom.dist(6, 10, 0.5, false) = 0.205.

If X = number of heads in 10 tosses, then

$P(X \leq 7)$ = binom.dist(7, 10, 0.5, true) = 0.9453

$P(X \geq 6)$ = 1 – binom.dist(5,10,0.5,true) = 0.3770

Betting on 7 in Craps Revisited

In the probability chapter, we discussed various gambling games and bets, including a bet that 7 would come up on the next roll of the dice in a craps game. As you may recall, the probability of winning this bet equals 1/6. If X = number of wins in n bets, then X has a binomial distribution with p = 1/6. Suppose a gambler makes 20 bets on 7. Then the probability that the gambler wins at least 15 of these bets is expressed as follows:

> P(X ≥ 15) = 1- binom.dist(14, 20, 1/6, true) = 0.00000001412, or about 1 in 71 million

Since the payoff odds for this bet are 4 to 1, in order to break even or make a profit, the gambler would need to win at least 4 bets out of 20:

> P(X ≥ 4) = 1 – binom.dist(3, 20, 1/6, true) = 0.43, less than a 50% chance, as you might have expected.

Mean and Standard Deviation of the Binomial Distribution

Just as we found the mean, or average winnings, for a gambling bet, we can find the mean, μ, or average value, of a binomial random variable, X, with n trials and with success probability p. The mean is the theoretical average number of successes in n trials.

The formula for the mean of a binomial variable is intuitive. For example, if you toss a coin 10 times, since the chance of heads on any toss equals 0.5, you expect to get 10 x 0.5 = 5 heads. If you toss a coin 50 times, you expect to get 50 x 0.5 = 25 heads. That doesn't mean that you will get exactly 25 heads every time you toss a coin 50 times, but if you toss a coin 50 times repeatedly, the average of the results will be close to 25. Random variation is always present, which keeps you from achieving precisely the mean every time you do an experiment, whether it's coin tossing or something else.

Consider the Internet ad example, in which the probability of a success (purchase) equals 0.2. It stands to reason that if you look at the result of 60 clicks on the ad, you will get about 20% successes, or 60 * 0.2 =12 purchases. In the dogs with fleas example, if you take a random sample of 80 dogs, the average is 80 * 0.4 = 32 dogs with fleas.

This leads us to the general formula:

> If X has a binomial distribution with parameters n and p, then $\mu = n*p$.

In a world with complicated formulas, this one's a breeze. The standard deviation (amount of variation) of a binomial random variable is straightforward.

> If X is binomial, with parameters n, p, then the standard deviation, $s.d.(X) = \sqrt{n*p*(1-p)}$.

The Geometric Distribution

The binomial distribution counts the number of successes in n independent trials with success probability equal p on each trail. The closely related geometric distribution counts the number of independent trials until (and including) the first success. For the geometric distribution there is no fixed sample

nazarovsergey/Shutterstock.com

size, because you keep going until the first success. For example, when repeatedly betting on 7 in craps, we used the binomial distribution to find probabilities for the number of wins in n bets. With the geometric distribution, we find the probability that the first win will come on the nth bet. For example, letting X = number of bets up to and including the first win, with p = 1/6, then $P(X = 5) = (5/6)^4 * 1/6 = 0.08$. The "5/6" term is because if your chance of winning this bet equals 1/6, then your chance of losing equals 5/6. The probability that your first win is on the fifth bet, P(X = 5), means that you lose the first 4 bets and then win. Since bets are independent, we use the multiplication rule for probabilities.

For the general formula, if you have independent trials with p = chance of success on any trial, and if X = number of trials until the first win, then X is a geometric random variable with

$$P(X=k) = (1-p)^{k-1} * p, \text{ for any positive integer } k.$$

When we analyzed roulette, we looked at the double-up strategy, in which you keep betting on red until you win, at which point you quit. The geometric distribution can be used to describe this strategy. The chance you win a red bet in roulette equals 18/38 and the chance you lose equals 20/38. What's the probability that you will lose the first 8 bets and then win? The number of bets until the first win, X, is a geometric random variable with p = 18/38 and 1-p = 20/38, so $P(X = 9) = (20/38)^8 * 18/38 = 0.002789$. Also, the probability that it will take at least 8 bets until you win, $P(X \geq 8)$, is equivalent to P(you lose the first seven bets), because if you lose the first seven bets, then it will take 8 or more bets until you win the first time. So, $P(X \geq 8) = (20/38)^7 = 0.0112$, or about 1 in 89. In other words, if you use the double-up strategy repeatedly, it will take you 8 or more bets until your first win about 1 out of every 89 times that you use the strategy.

What's the probability that you *never* win? The probability that it takes you at least k bets until your first win equals $(20/38)^{k-1}$ which is a fraction that gets smaller as k increases, so the chance that you never win must be smaller than $(20/38)^k$ for every k. The only number that satisfies this is 0, so P(you never win) = 0. That's heartening. You're guaranteed to eventually win! Unfortunately, you might go broke first.

Anton Veselov/Shutterstock.com

The Assembly Line

Suppose that 10% of the items coming off an assembly line are defective. You start sampling items. What's the probability that the first defective item is the 30th item you sample? Using the geometric distribution formula with p = 0.1, we find $P(X = 30) = 0.9^{29} * 0.1 = 0.0047$.

Mean of the Geometric Distribution

We can find the mean, or average, of the geometric distribution intuitively. In the previous example, 10% of the items coming off an assembly line were defective. Since an average of 10% are defective, it seems that you would find a defective item on roughly every tenth item that you sample. Theoretically, this turns out to be true. If X is geometric with success probability = p, then the mean of X = 1/p. This is true for any geometric distribution. When we bet on 7 in craps, then P(win) = 1/6, so the mean number of bets until we win for the first time equals 1/(1/6) = 6. If we count items coming off the assembly line until the first defective item, since about 10% of the items are defective, then the mean number of items until we see the first defective equals 1/0.10 = 10.

Bottom Line

Suppose you have a sequence of n independent trials, where each trial can be either a "success" or a "failure," and where the success probability, p, remains the same from trial to trial. Then the number of successes in the n trials has a binomial distribution with parameters n and p. This distribution has been well-known for hundreds of years and describes one of the most basic random processes. The probability distribution formula is based on the combinations formula and is easy to use with spreadsheet software and messy to use by hand or even with a calculator.

BIBLIOGRAPHY

Faux, Zeke – *Number Go Up, Inside Crypto's Wild Rise and Staggering Fall* - Currency, an imprint of Random House, 2023

Jung, C. G. *Synchronicity: An Acausal Connecting Principle* - (trans. R. F. C. Hull; Bollingen Ser.), Princeton, NJ: Princeton University Press.

Kahneman, Daniel, *Thinking, Fast and Slow,* Farrar, Straus, and Giroux, 2011

Mosteller, Frederick, and Diaconis, Persi, *Methods for Studying Coincidences,* Journal of the American Statistical Association, December 1989, Vol. 84, No. 408, Applications & Case Studies.

Orkin, Michael, *Probability and Statistics Made Interesting,* 2nd Edition, Kendall Hunt, 2022

Orkin, Michael, *Games of Chance and Games of Skill,* Chance Magazine, American Statistical Association, November, 2021

Orkin, Michael, *Can You Win? The Real Odds for Casino Gambling, Sports Betting, and Lotteries,* W.H. Freeman and Company, 1991

P.B. Stark and D.A. Freedman: "What is the Chance of an Earthquake?", Earthquake Science and Seismic Risk Reduction, v. 32, Kluwer, Dordrecht, The Netherlands, 201-213. September, 2001.

Taleb, Nissam Nicholas: *The Black Swan*, Random House, 2007.

INDEX

www.ingramcontent.com/pod-product-compliance
Lightning Source LLC
LaVergne TN
LVHW050816061126
851857LV00012B/833
* 9 7 9 8 3 8 5 1 2 2 5 2 3 *